T0061967

" Besides being beautiful little hand-sized objects themselves, showcasing exceptional writing, the wonder of these books is that they exist at all . . . Uniformly excellent, engaging, thought-provoking, and informative."

Jennifer Bort Yacovissi, *Washington Independent Review of Books*

" . . . edifying and entertaining . . . perfect for slipping in a pocket and pulling out when life is on hold."

Sarah Murdoch, *Toronto Star*

" Though short, at roughly 25,000 words apiece, these books are anything but slight."

Marina Benjamin, *New Statesman*

" [W]itty, thought-provoking, and poetic . . . These little books are a page-flipper's dream."

John Timpane, *The Philadelphia Inquirer*

" The joy of the series, of reading *Remote Control*, *Golf Ball*, *Driver's License*, *Drone*, *Silence*, *Glass*, *Refrigerator*, *Hotel*, and *Waste* . . . quick succession, lies in encountering the various turns through which each of their authors has been put by his or her object. . . . The object predominates, sits squarely center stage, directs the action. The object decides the genre, the chronology, and the limits of the study. Accordingly, the author has to take her cue from the *thing* she chose or that chose her. The result is a wonderfully uneven series of books, each one a *thing* unto itself."

Julian Yates, *Los Angeles Review of Books*

The Object Lessons series has a beautifully simple premise. Each book or essay centers on a specific object. This can be mundane or unexpected, humorous or politically timely. Whatever the subject, these descriptions reveal the rich worlds hidden under the surface of things."

Christine Ro, *Book Riot*

. . . a sensibility somewhere between Roland Barthes and Wes Anderson."

Simon Reynolds, author of *Retromania: Pop Culture's Addiction to Its Own Past*

OBJECTLESSONS

A book series about the hidden lives of ordinary things.

Series Editors:

Ian Bogost and Christopher Schaberg

Advisory Board:

Sara Ahmed, Jane Bennett, Jeffrey Jerome Cohen, Johanna Drucker, Raiford Guins, Graham Harman, renée hoogland, Pam Houston, Eileen Joy, Douglas Kahn, Daniel Miller, Esther Milne, Timothy Morton, Kathleen Stewart, Nigel Thrift, Rob Walker, Michele White.

In association with

LOYOLA UNIVERSITY NEW ORLEANS   Washington University in St. Louis

BOOKS IN THE SERIES

perfume

MEGAN VOLPERT

BLOOMSBURY ACADEMIC
NEW YORK • LONDON • OXFORD • NEW DELHI • SYDNEY

BLOOMSBURY ACADEMIC
Bloomsbury Publishing Inc
1385 Broadway, New York, NY 10018, USA
50 Bedford Square, London, WC1B 3DP, UK

BLOOMSBURY, BLOOMSBURY ACADEMIC and the Diana logo are trademarks
of Bloomsbury Publishing Plc

First published in the United States of America 2022

Bloomsbury Publishing Inc does not have any control over, or responsibility for,
any third-party websites referred to or in this book. All internet addresses given
in this book were correct at the time of going to press. The author and publisher
regret any inconvenience caused if addresses have changed or sites have
ceased to exist, but can accept no responsibility for any such changes.

Library of Congress Cataloging-in-Publication Data
Names: Volpert, Megan A., author.
Title: Perfume / Megan Volpert.
Description: New York: Bloomsbury Academic, 2022. | Series: Object lessons |
Includes bibliographical references and index. |
Summary: "A cultural and philosophical exploration of the art, science,
and business of perfume"– Provided by publisher.
Identifiers: LCCN 2021047849 (print) | LCCN 2021047850 (ebook) |
ISBN 9781501367144 (paperback) | ISBN 9781501367151 (epub) |
ISBN 9781501367168 (pdf)
Subjects: LCSH: Perfumes–History. | Perfumes–Social aspects. |
Perfumes industry–History.
Classification: LCC GT2340 .V65 2022 (print) | LCC GT2340 (ebook) |
DDC 391.6/3–dc23/eng/20211112
LC record available at https://lccn.loc.gov/2021047849
LC ebook record available at https://lccn.loc.gov/2021047850

ISBN: PB: 978-1-5013-6714-4
ePDF: 978-1-5013-6716-8
eBook: 978-1-5013-6715-1

Series: Object Lessons

Typeset by Deanta Global Publishing Services, Chennai, India
Printed and bound in the United States of America

To find out more about our authors and books visit www.bloomsbury.com
and sign up for our newsletters.

CONTENTS

INTRODUCTION

We could just say my father had a strong aversion to perfume, so the origin of my interest in it becomes obvious. Or we could skip over Freud to point a hippy dippy finger at the Zodiac, my Libra air sign. Determining an inception point for this passion is less important than sniffing out how it has become fundamentally a part of my character. Perfume has armed and disarmed me. As an object I deliberately and accidentally engage with many times each day, it is a gift to me and it structures my own giftedness as a spooky punk. Perfume and I suffer from a similar perception problem—that we are obnoxious, impenetrable, esoteric. It doesn't take an expert to fathom the complexity of fragrance, only a willing aspirant. I am here as an amateur aromachologist, making some philosophical observations from the middle of a life-long exploratory dabble in the impact of odors on the psyche.

Admit it: you yourself can conjure up exactly the way your house smelled when you were a kid. Or the waft of your favorite cookie in the oven, or the fumes of gasoline. So don't worry that you can't handle an object lesson in perfume.

Sure, you'll encounter some new words and some other uses for words you already know. Sometimes the ideas will be weird and fragmentary, other times they will be intuitive and smooth. Because that's life, and each of us must follow our own nose to navigate it.

Consider your confidently advanced and nuanced discernment of the scents of popcorn. You know the dazzling slickness of movie theater yellow and too much saline. But there's also the other one: its faded glory snaking in low from the workhorse across the hall, slightly burnt but probably salvageable, oddly captivating, comforting even as it annoys. Remember that you have developed a strategy that tries to strike a balance between these two smells as you stand at the microwave with your own un-popped bag.

Late one afternoon at the high school, my news nerds and I are rushing to meet deadline over a crumpled mountain of popcorn bags. A student raises her hand and I head over to her workstation. She is among the school's best, in that high class suburban way of knowing nothing about life except how to succeed at it in the same blinkered way as a racehorse does. When I arrive at the edge of her personal space, the zephyr of my movement across the room advances ahead and she recoils, scrunches up her nose and narrows her eyes. "You smell like my grandfather," she comments. Her tone is more neutral than what her face betrays, but I pick up a few heads turning in my peripheral. They have come to recognize her passive aggressions and most of them wearily capitulate to her strong but fledging will to power. I bend to look at what's on her screen, see why she called me over. The sweat in her hair from gym class two hours ago has reactivated the scent of a costly spa-brand shampoo. Responding in the same volume as her commentary, I say, "Then your grandfather has somewhat expensive taste."

Satisfied by the slight sting of "somewhat," everyone returns peaceably to work. The minutes-long scene they'd hoped for, in which I read this child for filth for any reason at all, doesn't materialize. After we solve her problem, I nonchalantly touch a hand to the side of my face to bring my wrist close to my nose, quietly and covertly recentering myself by inhaling a whiff of the fragrance that there is almost no chance her grandfather shares with me, *Aoud Forest* by Montale. It's mostly a base note now. The top notes of grapefruit, lemon and rosemary that first greeted me this morning having burnt off within a half hour of that initial spray, the heart note of ginger fleeing around lunchtime but its rose companion still a haunting whisper. The base notes, those tougher, fixative molecules that should still be available for my wife to smell when I get home from work, are white musk and ambergris. Ambergris is a secretion of the bile duct in some sperm whales so that their bellies can repair after ingesting the shards of a squid's beak. It is literally either vomited or pooped out by whales, then it floats near a coast and a perfumer snatches it up. Even though the musk of ambergris mimics the earthy sweetness of oud, *Aoud Forest* isn't technically an oud because oud comes from agar trees. When the tree gets infected by a specific type of parasitic mold, it produces a resin that is harvested for fragrance. The rarity and exoticism of ambergris and oud are a large part of their mystique in the perfume industry.

The real deal ambergris is no longer actually in use anywhere. My perfume uses a cheap synthetic approximation and yet a 100ml bottle of Montale costs $170. I used to fend for myself with not much money. Now I collect fragrances, which is generally considered a very snobby hobby, like being a wine person. One can calculate cost per milliliter of perfume just like cost per glass of wine and then decide at what point the expense becomes unreasonable. For example, a 50ml bottle of Tom Ford's *Lavender Extreme* costs $335. That's too expensive for me, but I am working off a sample provided by my girl Simone at the Tom Ford counter in the basement of Bergdorf's. I wanted to wear it while writing this introduction because it had an appealing marketing campaign promising a loudly deconstructed synthetic, something perhaps akin to the mood of this book. Other than the price, *Lavender Extreme* turns out to be not very extreme. It dries quickly down to a vanilla base note, utterly fine and palatable. Maybe a little extra clean, rather than merely creamy. That's about quality of ingredients, rather than the mix, because you can mix lavender and vanilla at Bath and Body Works prices. I'm wearing a soapy vanilla, not an edible one.

If we are what we eat, we might do well to remember that what we taste is actually around seventy percent smell. After I'm finished writing today, there's overnight oats made with blueberries and lemon curd awaiting me in the fridge, and it will be set off nicely by *Lavender Extreme*. The vanilla base note will soften up the tartness of the lemon curd and whatever is left of the lavender heart note will strengthen the juicy indigo impression of the blueberries. You can radically alter a dining experience by fragrance, and this is part of why smoke is such an essential ingredient. By accident of the necessity of heating our food in order to cook it, smoke is probably the first human-made fragrance.

In New Orleans, Smoke Perfume is a one-woman operation working with small batches of essential oils made from local resources like cypress and satsuma. She's got one heck of a rose grower by the name of Miss Jeanette. The scent of these roses performs a spicy, warming sensation within the body. The magic is so precise. Roses are an acquired taste for a certain kind of person, like wearing the color pink. So much baggage there, it took me years to get into an experiment with either.

A scent is so tenuous, hanging there for just a moment in the ether. This is what makes the spell of smell so tenacious. Perfume always only evokes. It is ever behaving as a tether to other things, a tool of connectivity. Colors, attitude, memory. It's incredible that we can bottle that. And sell it. Scent is free; perfume is not. And there are a million kinds of vanilla. For a solid object—and a sold object—perfume is intangible to a shocking degree. This is why it's an excellent gateway to profundity at large. The airspace of our interrogation is limitless and the presence of some answers in no way suggests the absence of many further questions. Moving toward achievement of this awareness is fundamentally what it means to live a more fragrant life. Recognize that you are already having many daily thoughts and feelings about the smells you encounter and that you often make an action plan based on those. Now let's get funky.

1 SCIENCE

Perfume is a colorless green idea that sleeps furiously. That's a Chomsky joke. "Colorless" and "green" are contradictory, yet an object such as a green glass window or a soap bubble can simultaneously capture both ideas anyway. Sleep is a verb, a nesting doll hosting presumptively good choices for adverbs such as "calmly" and "peacefully," and so sleep appears to inherently reject the prospect of a furious state. Yet one might thrash around in the night or dream very lucidly, sleeping in a manner that is indeed furious. The sentence itself, this assertion about perfume, is perfectly cogent.

In 1957, Noam Chomsky noticed that a sentence conveying utter nonsense can remain grammatically correct. His example: "colorless green ideas sleep furiously." What a gorgeous sentence. Our lived experience of "green" or "sleep" cannot quite align with our ostensibly objective definition of those words. According to this transformational theory of grammar, surrealism is alive and well. Paradoxical things are imbued with vitality not in spite of our language rules, but because of the very verve of them.

Perfume has always seemed to me to be exactly this type of oddity. Our brains do a magic trick with smells that is much like the one they do with words: in the jumble of nouns and adjectives and other parts of speech, our brain looks for patterns in word order that help to build a meaning. Sometimes a very classic and rigidly comprehensible structure like Chomksy's example sentence turns out to generate the wildest bit of pure poetry.

A violet is too delicate to undergo scent extraction directly from its raw natural material, but fortunately, nail polish remover (acetone) plus lemongrass (citral) divided by battery acid (sulfuric acid) equals the smell of violets (beta-ionone). That's actually science, a chemist's ability to mix molecules together until they form a structure the nose finds indistinguishable from the complex natural formula of a violet.

Specific clumps of molecules are the building blocks of perfume. The nose delivers these to our brain, which then tries to make sense of them. The brain recognizes a flower molecule it received, as well as a molecule that is smoky and animalic. Maybe the animal is unspecific. Consider the extent to which blindfolded people might have difficulty identifying a cow versus a horse by smell alone. So the information from these molecules up one's nose is freely associating several concepts one's brain has stored from past experience, bringing a few of them together in a totally new relationship, a hallucination that is the scent equivalent of language syntax. The brain organizes these molecules into the best meaning it can: a rose made of leather, clear as day, though our eyes have never seen it in material form. We have seen it in scent, sleeping furiously in its green and colorless way across our mind's eye.

We know a fair amount about how the mind reads words and shockingly little about how it translates smells. And yet, smells are just molecules, and we know almost everything about molecules. We even know that only five types of atoms are typically involved in smelly molecules: carbon, hydrogen, oxygen, nitrogen and sulfur. These atoms connect together into molecules, and after those molecules fly up one's nose, science doesn't have much to say about why one's brain reports back with images like the rose made of leather. The rose and leather scents are not objective properties of a molecule just like color is not an objective property of light wavelengths. The challenge lies in figuring out how our biology recognizes these signs from outside itself. A smell is a kind of feeling we have when the cells inside our nose touch certain molecules.

These cells are smell receptors, discovered in 1991 by Linda Buck. A molecule is like a puzzle piece that locks into a receptor. It might be like playing with a kid's toy, where one can't shove the triangle block into the square hole. There's also the question of chirality, or mirror-image molecules, which is like asking whether the triangle block will fit into the triangle hole three different ways as one rotates it. Perhaps a receptor could pick up a smelly molecule and its opposite, except that the structural oppositeness of two molecules doesn't correspond to their smells. Neither does structural similarity.

An awesome thing about the nose is that it's basically a helmet covering the outermost portion of one's brain. There are hundreds of smell receptors covering millions of olfactory neurons. This clump of neurons is known as the olfactory bulb. This bulb is the only part of the brain that pokes out through one's skull, blowing in the breeze of one's nostrils and gathering the backdraft from one's mouth. The bulb collects info about the smelly molecules and sends this info to parts of the brain governing sensory perception but also to the parts governing memory and emotion, including the amygdala and the hippocampus.

The amygdala, responsible for one's fear and anger, is usually characterized as the reptile part of the brain. It bolsters our survival instincts or what psychologists call the id. The hippocampus is responsible for executive functions like learning, mapping and remembering. So perfume not only immediately touches the literal brain, but also offers information that proliferates more broadly across multiple segments of the brain in ways that other sensory inputs like touching or seeing do not. Rodents have about a thousand kinds of smell receptors, while humans have less than half that number. With our measly four hundred receptors, humans have an easier time of identifying differences between two smells rather than what they have in common. It's commonly understood that human noses don't know as much as those of other animals.

Perhaps this is why science has left so much about smell unstudied. We already know we don't get top marks compared to a lot of other species, so the entire faculty of smelling is demoted as less than civilized. We invent things to close the gap, for example carbon monoxide detectors to sniff out the deadly gas that could otherwise poison us. Aromachology and aromatherapy are denigrated as less than rigorously data-driven, as bunk science despite their two thousand or so years of contribution not only to survival but to civilization. Inventions like Little Tree car air fresheners or Febreze odor eliminating spray may not seem like high art, but they are not useless or dismissible creations.

There's a television commercial for odor-eating Febreze featuring a very visibly stinky apartment full of piles of trash and dirty socks and so on. After some Febreze is sprayed into the room, people are led in blindfolded to report that it smells wonderfully fresh and clean. Take off the blindfold and the eye would disagree with the nose. It's a contradiction that proves the product will work to mask unpleasant smells. Whether covering the smell of dirty underclothes is a worthwhile scientific mission is a matter of opinion, but the profit potential for a product resulting from such a mission is undeniable. So smell science continues to gravitate toward types of research that are most likely to make money.

This begs the question of how much one would pay for an improved sense of smell. There are a variety of medical conditions describing impacts on the capability of one's nose that fall under the general heading of parosmia: inability to detect a full range of scents. Cacosmia: detecting an unpleasant odor around everyday things that do not actually possess an unpleasant odor, for example due to Alzheimer's, Parkinson's or schizophrenia. Phantosmia: hallucinating "ghost" scents where none are actually present, for example due to traumatic brain injury, epilepsy or multiple sclerosis. Hyperosmia: heightened and usually overwhelming sensitivity to odors, for example due to hormones, genetics or migraines. All culminating in anosmia: an inability to smell.

The consequences of such conditions are as quantifiably vague as they are qualitatively obvious. One needs to be able to smell the leftovers in the fridge to know whether they've gone bad or can still be eaten. This is a matter of survival, yes, but also a bedrock of the enjoyment we take in the world. Imagine never knowing what a pizza smells like. Or perhaps worse, imagine having once known what a pizza smells like but now it smells like literal poop, or it can't be smelled at all. Parosmia of any kind can radically reduce one's quality of life by limiting feelings of belonging, and the social isolation can be depressing or debilitating. Imagine not knowing whether one needs to hop in the shower, or where exactly the dog peed on the carpet again, or how one's house is permeated with the scents of a big family dinner at Thanksgiving.

It's estimated that more than half of these conditions are due either to head trauma or to upper respiratory tract infection. Now is a good time to say that this book was written almost entirely during the COVID-19 pandemic.

Early in spring of 2019, scientists told us that losing one's sense of smell was a common symptom of coronavirus. A year later, we would learn that this absence often persists among so-called long-haulers as one of the most pernicious lingering consequences. Because I had to go into a packed classroom to do my essential work as a teacher at a time well before the vaccine, when the nature and extent of transmissibility was mostly unknown, I scented my face masks with violet and mint to help stave off panic attacks about health and safety. Bacon-scented face masks were a fad for a while. Suddenly everyone was aware of not only each other's physical space for social distancing, but also the basics of aerobiology—droplets, aerosols, the lack of borders or privacy in our airspace. Dogs trained to sniff out ovarian cancer or tuberculosis were retrained and deployed in the Helsinki airport as an experimental alternative to rapid coronavirus testing. Noses were top of mind for once. My own nose was among the thousand things I worried about, along with whether the research for this project would be compromised if I caught the dreaded virus.

I was once temporarily anosmic and trying to remember how it felt still creeps me out bigtime. Some years ago, after a particularly overindulgent field trip to the fragrance counters of a major department store, I ended up with a bad migraine that cost me my entire sense of smell for about forty-eight hours. During that horrifying but thankfully brief window of time, I ate a lavish home-cooked meal that a friend had spent several hours preparing. All I can say about it is that most of the food was warm, there was a variety of textures, and it all looked great on the plate. When we believe we are tasting something, about seventy percent of the data our brain receives is actually based on smell. This is why pinching one's nose helps bitter medicine to go down. To this day, the chef doesn't know I couldn't smell or subsequently really taste any of it, as if this revelation would somehow have been an embarrassment to both of us. And I remain sad that a terrific meal was lost upon me while I was transitorily dysfunctional.

As an antidote to my fear that if I lost my sense of smell I'd be unable to complete this book, I bore in mind that the best restaurant in America was run by a chef who had lost his sense of smell. Grant Achatz, of the Chicago molecular gastronomy temple Alinea, was diagnosed with stage 4 cancer on his tongue in 2007. He thought he would have to cut it out, thereby forever surrendering most of his ability to smell and taste as well as imperiling his reputation as a rising culinary star. Instead, he went in for several months of aggressive radiation and chemotherapy, and he was cured without the major surgery. Yet he continued to design highly experimental and precise menus throughout the window of time where he had no genuine ability to evaluate the foundational aspects of his own compositions. It led him to new ways of thinking about color, texture and other elements of food preparation, as well as strengthening bonds between the team members whose judgment was necessarily subbed in for his own. Chef Achatz got to keep his tongue and eventually rebounded from his anosmia.

February 27th is Anosmia Awareness Day. Because the condition is invisible, many people underestimate the extent to which it disables. Anhedonia: the inability to experience pleasure for activities that are usually found enjoyable, for example trying new restaurants, cuddling with someone who is wearing perfume or going for a hike in the woods. Imagine loss of appetite, loss of romance, loss of peace of mind. So much do we forget how our noses make adventure, so much do we take our sense of smell for granted, that it can be challenging to work up empathy for those who suffer from parosmia. It takes a village of neurologists, physiologists, biologists, geneticists, chemists, and psychologists working in their diverse disciplines to weave together some semblance of comprehension and compassion for parosmics. The world's only independent non-profit institute for advancing discovery in taste and smell is the Monell Center in Philadelphia, which houses about fifty interdisciplinary scientists. There are also a few small groups focusing on publicity, education and resources like the Anosmia Foundation in Canada and Fifth Sense in the United Kingdom.

The world lacks a celebrity spokesperson for the cause. The English Romantic poet, William Wordsworth, had no sense of smell. Actor Bill Pullman lost his sense of smell after an accident during a stage performance where he fell fifteen feet, hit his head, and was in a coma for several days. He sometimes does advocacy work with the Monell Center. Ben Cohen, of Ben & Jerry's Ice Cream, has no sense of smell. His corresponding lack of tasting ability caused him to focus on the texture of ice cream more than the flavor, and this is why the vast majority of the brand's products feature big chunks and syrup swirls. Cohen doesn't like for any spoonful to be simply smooth.

If we are what we eat and seventy percent of eating is engaging with smells, then it is fairer to say that we are what we smell. The fact that most of eating and drinking is about fragrance rather than flavor generates some intriguing philosophical propositions. Consider whether one can taste something by only smelling it. This runs the gamut from pizza to sewage, the question of whether one is tasting these each time one walks by a pizza parlor or a public restroom. And smells are of course not uniformly perceived. Roses called by other names may not deliver the same sweet scent, in the same way red velvet cake doesn't seem to taste like chocolate cake. These two cakes are ninety percent identical in the practical realities of baking and taking a bite, yet their symbolic relationship is as dumbly fraught as the relationship between chocolate and vanilla. Chocolate is composed of about seventy percent vanilla, so spending too much time on parsing the allegedly objective aspects of flavor or fragrance turns out to be an absurd enterprise. Chocolate and vanilla are characterized as definitional opposites, yet chocolate is living the paradox of being made mostly of vanilla.

The rational failure of this metaphor is oddly sensible in practice, like the surrealism of Chomsky's colorless green ideas, and it's easy to make the leap from food to perfume.

Scientific principles—like experiment, trial and error, hypothesis, adjustment based on data, final product—apply to cooking and thereby also to food criticism or restaurant reviews. I came by my own scent vocabulary very much through the language of food. This language is accessible to anyone because everybody sniffs in the same way that everybody eats. Fragrance is as much a daily inevitability as flavor. We used to use perfume internally as well as externally and the basic plants-water-alcohol nature of perfume sits midway along the spectrum between a cup of tea and a cocktail. Modern tea masters and sommeliers would agree that most food science concepts have some parallel in perfumery and one's background knowledge of food can easily be the launchpad to a love of perfume.

Jeffrey Steingarten, also known as the meanest judge from the Food Network's *Iron Chef America* reality television competition show, often uses scent as the entry point for his critiques. He's always leaning over plates for a big inhale and did once turn his attention directly to perfume in the pages of *Vogue* magazine. This was to amusingly denigrate a moment in the early Nineties when pheromones became all the rage because a California company by the name of Erox released a pair of fragrances called *Realm*. Most animals communicate with the chemical scent trails known as pheromones because they don't have other language. There is not overmuch evidence to suggest humans relate to each other based on pheromones, other than the possibility that it causes menstrual cycle syncing in women's college dormitories. Nevertheless, *Realm* offered a fruity floral and an aromatic, fern-like fougère that vaguely claimed to lure the opposite sex by means of science.

Twenty years of meager market share later, the internet yields not one single anecdote from anybody attributing their happy coupledom to *Realm*. The merchandise in the Clovers' 1959 hit, "Love Potion No. 9," remains mythic. There is not yet any reasonable evidence that the vomeronasal organ wedged inside our noses performs the pheromone reception functions it performs in other animals. A corresponding myth is that humans have a poor sense of smell compared to other animals. It's true we have fewer receptors, but we are about average with other large mammals. Our brains also have more robust computation capacity for interpreting odors than many other animals do. The idea that humans are weak in the nose proliferated during the nineteenth century, when neuroanatomist Paul Broca cut open a skull, measured our organs, and declared that humans have proportionally smaller olfactory bulbs than a lot of other species. When Sigmund Freud picked up on this one guy's supposition that surely size matters, lack of sniffing skills got bundled into the total package of human anxiety and it threw a damper on proper scientific exploration of our sense of smell for a century.

Yet perhaps our lived experience alongside other animals makes Broca's unsubstantiated conclusions feel right. Don't wear perfume on a hunt because it spooks the deer. Don't leave a single crumb in the darkened corner of a cabinet because a mouse will surely sniff it out. Sharks can smell their food underwater. Bears can smell fear.

The most common data we collect in everyday life pertains to dogs: our companion animals are always snuffling at something or other, picking up on signals we don't think we can detect. A well-trained cadaver dog can catch the scent of a sliver of human remains from miles away and track it to the source, though it may take a while and the evidence may not hold up in court on the authority of the canine alone.

An equally well-trained human would give the dog a good run for its money, but most people don't bother to practice their sense of smell enough to get there. Instead, nowadays, paying so much heed to the nose might fall under the heading of mindfulness. Stopping to smell the roses has more potential than a mere boost to one's mood. The more examples of "rose" one smells, the more the nuances of each become clear and the more expansively diverse one's scent vocabulary becomes. This is where the rubber of science meets the road of spirituality.

Iconic disability rights activist Helen Keller reflected that she understood the souls of objects and other people based on the properties of their scents. Facing the challenge of being both deaf and blind, Keller exhaustively cultivated her sense of smell and judged anyone who neglected to exercise this faculty as being willfully ignorant. She longed for an Einstein to piece together a proper science of scent and argued that this still relatively elusive knowledge would evolve humankind toward the infinite.

Meanwhile, perfumers have developed scientific processes for us to practice. Jean Carles, the French perfumer whose creations included such icons as *Ma Griffe* for Carven, *Miss Dior* for Christian Dior and *Canoe* for Dana, founded the Roure Perfumery School in 1946. The methodology that bears his name is still in use today. It's a spreadsheet where each row classifies the building-block scents of perfumery, both synthetic and natural, into a family, and each column offers a different example from that family. The citrus family row contains lemon, bergamot, mandarin, orange, lime, then finally grapefruit. One sniffs across a row to feel out the similarities as if they were all varying shades of the same color. One sniffs down a column to cover more than one family, from a citrus to camphorated to green note for example, yielding an immediate feeling of differences between lemon to lavender to galbanum as if looking at yellow then purple then green. With practice, these end up memorized. I sometimes play a lowkey version of this game when I'm sniffing around perfume counters by giving myself constraints. For example, I'll only pick up perfumes in blue bottles and check out what kind of fresh, aromatic water energies each possesses. Or I'll only pick up perfumes that include an amber accord and check out how many variations on this theme there are. To study perfume is to go on adventure and receive surprise. It's a twisted mountain trail with jolts of consternation and revelation hidden around every unusual turn, not a straight-away on flat terrain with good visibility where one is free to speed. Realize that the temptation to reach for a map is a sign of one's overreliance on objectivity.

A certain amount of surrealist bemusement or cognitive dissonance is required if one is to travel a life-long path of learning about perfume. The science is rather full of gray areas and we shall soon discover how the literature of it is—ironically—often quite black and white. But part of the beauty of perfume, to carry through the metaphor of Chomsky's transformational grammar, is that it operates like a preposition. Perfume goes on a person, or into a bottle, or through the air. It's a builder of experiential bridges and connector of ideological networks, performing as a connective thread across all that is known inside an individual's mind. A fragrance weaves its way freely through associations, images and even sounds, bringing a few of them together in totally new inventions and previously unknown relationships more suited to the mode of poetry than cartography. This is also the definition of synesthesia: the perceptual phenomenon wherein stimulation of one sensory or cognitive pathway, such as smell, leads to involuntary experiences along additional pathways, such as sight or sound. It's possible to constantly reorient oneself to perfumes by pairing them with other forms of art or science with which we feel more familiar. One can pair a playlist with a fragrance just as one pairs a wine with dinner.

A set of drums or paints or flower essences are equal tools. These forms or disciplines or pathways are all equally suited for reflection and conversation, or they ought to be.

So sayeth a highly decorated French New Yorker, chemist, and master perfumer Christophe Laudameil in his excellent little manifesto full of concrete suggestions, "Liberté, Égalité, Fragrancité." He would like to see three things happen in the world of fragrance. First, the public must be better educated, for example given more access to olfaction information from university courses to creative career pathways. There are myths to dispel and crap products to weed out. We must be more discerning and less fearful. Second, the business side must level up, for example committing to ethics codes for intellectual property and sustainable development. There are heroes working in the field of scent. We need fresh waves of inventiveness. Third, the total effect of changes in education and industry must ultimately point toward a more fragrant manner of everyday living. A chef and a perfumer working in the same lab, for example. We are already materially able to carry out these dreams of colorless green, even if on paper they sometimes come across as too weird, sleeping furiously until more people open their nostrils.

2 LITERATURE

Perfume always goes to character and characters in the orbit of particularly fragrant symbolism are most often found in the genres of romance and mystery. Allison Breed uses her Twitter account, @RomanceSmells, to serve up an ongoing project called the *Male Scent Catalogue*, in which she keeps a running list of ways men's odors are described in the romance novels she voraciously consumes. The result is a delightfully predictable combination of mockery and celebration, as there is apparently no end to the number of ways one can say a man smells both clean and like the woods. These descriptions often come in threes, a top-heart-base configuration that always seems to end on the essence of manliness. In other words: "he smelled like himself." A majority of romance novels fail to get beyond the bunk science of pheromones and the bunk advertising of masculinity, resulting in this unthinkingly circular proposition that men smell like men.

By comparison, mystery novels offer a more nuanced perspective in which perfumes are one intriguing tool in the box of dark communication strategies. Beyond seduction, by "dark communication" I mean that a perfume can create avenues for deception, subterfuge, transgression, deviance and so on. Napoleon is said to have gone through more than 100ml of his lime and rosemary cologne in baths and dabs every day. Copies of this invigorating scent are still a best-selling formula and the political writings of a Renaissance diplomat, the Italian philosopher Niccolò Machiavelli, help explain why. In his 1532 treatise *The Prince*, he famously argued that one's benevolent ends may justify the use of any necessarily fraudulent or violent means, and that leaders ultimately do better to be feared than loved. Napoleon seemed cool for a short while because an elected official utilizing dark communication is presumably off the hook when it helps achieve big benefits for the citizenry. Or at least big profits for the politician's backers. But in the land of mystery novels, the agents of a bad action almost always get foiled or caught. Their goals are seldom validated as righteous, and it is most often the villains who are mixed up with perfume.

Take the latest book by best-selling Canadian mystery writer Louise Penny, for example. *All the Devils are Here* is the sixteenth installment in a series she began publishing in 2006, wherein Chief Inspector Armand Gamache of the Quebec Provincial Police always succeeds in finding out who the murderer is and why they did it. Penny pours on the perfume in a variety of meaningful ways, situating Gamache in Paris for the birth of his grandchild when his godfather is nearly killed by a suspicious hit and run. Sixty pages in, readers are introduced to a highly intelligent engineer who reeks of Dior's *Sauvage*. This archetypal aromatic fougère for men beats one over the head with pepper and lavender, a loud blue smell repped in the ad campaign by Johnny Depp, whose own real-life characterization seems to grow more villainous all the time. Dior profits by Depp's bad boy image. This woman in the story may or may not be a mole, double-agenting on behalf of the bad guys, a possibility that opens in the moment she is portrayed simply as a woman who wears men's cologne.

Cologne is also a literal clue in the story. Upon finding the freshly dead body of a stranger in his godfather's apartment, Gamache and his wife suddenly realize they've stumbled into imminent danger because the murder's cologne still lingers in the air. It's a citrus scent, so the trail must be fresh because citrus notes don't tend to linger, leading the couple to conclude the villain is still in the apartment with them. Later, the wife explores a department store perfume counter in search of the murderer's cologne. It turns out to be *4711*, a slightly herbaceous, lemony fragrance produced since 1792 in the city of Cologne itself. Two or three suspects in the novel may be wearing it, repeatedly producing the dark communication conditions for mistaken identity and red herrings, and the Gamaches must literally sniff out the killer. Echoing the presumption of a romance novel, Inspector Gamache himself smells merely of a brandless and unspecific sandalwood. The hero doesn't do product placement.

As the mystery unfolds, it's clear our hero is no expert in matters of fragrance. The good guys rarely are. This is most in evidence in the quintessential character of Detective Sherlock Holmes, brainchild of Sir Arthur Conan Doyle. Holmes is always portrayed as having an acute sense of smell and a widely catholic interest in smells. Using just his nose, he might trace an object left at the crime scene to the one and only dock where the criminal's lair is hidden, or he might distinguish between a variety of suspects by the scent of their shoe polish. He always hits it rather quickly on the nose. Yet there is something about Sherlock that leaves a reader uneasy. Subsequent film and television portrayals of him have leaned heavily on his interest in fragrance as a mark of his intellectual giftedness and equally a sign of his darkness.

He is the kind of protagonist who is not very likable, being either melancholy or rebellious most of the time—a Byronic hero, named for the English Romantic poet Lord Byron. Holmes is classically anti-heroic in this vein, and still we root for him to solve mysteries with his no less mysterious power of deductions formed from trace evidence. He literally pulls a fragrant clue from thin air, all the while berating his naïve wing man Watson or the bumbling officers of Scotland Yard in his curmudgeonly way, unraveling the crime methodically as only a prodigy can, reliant on the simultaneous talent and curse that is his superior ability to apprehend and comprehend smells. His astuteness is its own burden, and he knows it—and dulls it in his off hours through drugs or sex or making a mess of his apartment or playing the violin. Sherlock is a volatile bohemian, but he has benevolent ends, solving crimes while remaining isolated unto himself as a tortured individual. The police overlook his ill-temper because they need his help.

In an amusing reversal, the character of Sherlock has proven good fodder for perfumers. It's common to find fragrancers creating intricate backstories to set the stage of a specific perfume. Imaginary Authors is a brand predicated on this intensively narrative style of marketing, and Sucreabeille is an example of a brand that offers scents for the fans of particular stories. Their collections include thirteen scents based on Joss Whedon's cult classic television show *Firefly*, and a series of scents based on the beloved sitcom *Golden Girls* with one named for each of the four protagonists and a fifth called *Thank You for Being a Friend*, after the show's theme song. Sherlock is ripe for interpretation in perfume, with most homages anchored by the smell of pipe tobacco and leather. There are online fan blogs devoted to tracking Sherlock's attention to smells in various stories, as well as others delivering speculative fiction on what fragrances the character might wear.

In the BBC adaption *Sherlock*, the perfume worn by his Achilles heel is not a matter of conjecture. Irene Adler, the cunning femme fatale who haunted Sherlock through many stories, wore Chopard's *Casmir*, an "oriental" vanilla with a very peachy top note that launched in 1992. From here, we could slide gracefully down a slippery slope into all the literary bad girls, from Old Testament Eve giving Adam a whiff of the apple to movie legend Elizabeth Taylor, who likely made as much money from her *Passion* and *White*

Diamonds fragrance lines as from her blockbuster films like *Cat on a Hot Tin Roof* and *Cleopatra*. Such a deep dive would be a gay old time, too, with dips into the poetry of Charles Baudelaire's *Fleurs du mal*—a primary source of inspiration often cited by especially French perfumers—or Oscar Wilde's costly libertine adventurism in *The Picture of Dorian Gray*. But all this examination of fallen women and queer prisoners will pile up into a garbage heap of stinky moralizing, so in support of the fine tradition of dark communication's deceits, let's not and say we did. As an alleged bad girl myself, I'd rather take this opportunity to give the world's most beloved novel about perfume a kick in the shins.

One cannot simply prattle on in praise of Patrick Süskind's *Perfume: The Story of a Murderer* and never bother to zoom out on the reasons why this book haunted bestseller lists for nearly a decade. With over twenty million copies sold it is one of the most profitable German novels of the twentieth century. Hollywood even picked it up and turned it into a blockbuster movie, because it offers such an easy moral judgment against dark communications: beware the dangerous seducer, avoid the fragrant deceiver, keep on your toes against the intellectual schemer. Its protagonist is a serial killer, gifted with a peerless nose but shackled in a scentless body, who makes a sublime perfume out of young virgin girls. He is eventually caught, whereupon by means of his most powerful perfume yet, an orgy ensues instead of his execution. There's also some cannibalism. Very heady stuff in this German novel from 1985—the kind of stuff that can only be answered by an offer of context provided by German philosophy from 1885. Simply put, Nietzsche's *Zarathustra* is a better novel than Süskind's *Perfume* because it succeeds in offering a mystic symbolism for perfume without all the attendant moralizing.

Among the fifteen works of philosophy that Nietzsche published during his lifetime, only number eight is a novel. *Thus Spake Zarathustra: A Book for All and None* is a fictional recasting of the life of the founder of Zoroastrianism, one of the world's oldest religions. Nietzsche depicted Zarathustra as the priest who bestowed upon humans a dualistic mortality pitting good against evil and who subsequently regretted what people did with that belief system; after a long isolation to contemplate how humans were using religion as a reason to kill one another, Zarathustra came back down the mountain to try to remedy what he had wrought.

The philosopher always considered *Zarathustra* to be his magnum opus and although perfume does not feature heavily in its plot the way it does in mystery novels, Nietzsche often makes quick references to scent because he firmly believed that sense of smell is a better metaphor for philosophy than sense of sight. Until Nietzsche, the art of deep thinking was mainly symbolized by the eye rather than the nose. Plato is responsible for this because of his own short story about the cave. In the allegory of the cave, folks held prisoner there can only see shadows on the wall and mistake them for reality. These pale imitations of actual life are made by a select intelligent few holding up objects in front of a fire—the world's great visual artists. If any prisoner escapes, they will find that real life outside the cave is far superior to the ignorant one inside the cave. But they would likely be torn to pieces if they try to go back into the cave and liberate others because the cave's darkness would leave them stumbling around blindly again.

Focusing on another product of the fire instead of the shadow, fellow Greek Heraclitus turns this story on its head with a trite dismissal, arguing that if everything we perceived were smoke, the nose would be our organ of judgment instead of the eye. Yet Plato's metaphor has endured as one is quite likely to believe the things one sees.

This is the whole danger for Nietzsche. For him, the cave reeks. The violently moralistic belief in good versus evil perpetrated by priests is the same as the shadows on the wall perpetrated by artists. He thought that ignorant people poisoned by ideology from the inside out could literally be diagnosed by their sickening stench, that the nose would not be deceived like the eye. This is in agreement with Heraclitus and an apt reversal of a common phrase: there must be smoke where there's fire. He once criticized Arthur Schopenhauer's philosophy by referring to it as a cadaverous perfume. It will come as no surprise that the revolutionary and free-thinking Nietzsche, who couldn't stand this bad air of largely foolish human company, could often be found hiking in the Alps. His smell was probably on the mountain man continuum so beloved by romance novelists.

A priest and a philosopher are only a breath apart, and in reflecting on why humans wag the dog—employ their ideas to make misery for other humans in a real war for a selfish good against a faked evil in a manner Machiavelli would have approved and Süskind's serial killer unequivocally implemented—*Zarathustra* is a novel that tackles perhaps the greatest mystery story of our existence and the genuine subject of all literature. Nietzsche wanted us to smell the lie as lie, to rely on scent as a more efficacious metaphor for thinking about thinking. A stench cannot be ignored, whereas one can easily turn a blind eye to shadows on the wall of a cave. Rather than craft the black and white box of a morality tale where villains are reliably brought to justice at the end by heroes who smell of sandalwood, Nietzsche constructed the perfume of *Zarathustra* out of three complicated accords, or in literary terms, themes. These are the overman, the will to power, and eternal recurrence.

Overman—or *ubermensch* or superman—refers to the idea that homo sapiens are not the end of the evolutionary line from apes, and that instead of thinking of one's highest ascent as a soul in heaven, one should keep both feet faithfully on the ground and continue to evolve on earth. *Zarathustra* implores us to get out of the way of bad odors and hold our nose until we can get into the open air. The anti-hero of the story is concerned that priests poison one's body with good/ mind evil/body dualism, the stench of which causes one's longing for the pleasantry of an afterworld in order to escape the decaying body. Nietzsche redirects this longing from a kind of place he doesn't think exists to a kind of person who may exist in the future. Humanity is at present like heavy drops of rain, whereas the overman is the lightning that waits beyond. Lightning is reported to smell like burning wires or plastic, ozone or chlorine, even cleaning fluid, by the way—inorganic materials.

One is a stinky idolater until overcoming one's reliance on the priests, politicians and detectives who all engage in poisonous forms of dualistic moralizing. This process of self-overcoming is a piece of the will to power. Nietzsche was fascinated by the scientific prospect that all living things have a will to live and he upgraded it to get beyond biological evolution into the territory of psychological growth. He offered a theory of how one derives pleasure from life. Cruel people derive pleasure from exercising control over others, as in the case of Süskind's serial killer sprinkling an angry lynch mob with a perfume that magically turns things around for him. The killer hurried to become a master, lest as a scentless outcast he inadvertently become a slave—again, that's good old dualism at work, yet also the reason Nietzsche's work ironically ended up working for Hitler.

But *Zarathustra* preaches the gospel of self-perfection and says happiness should smell of the earth instead of smelling like contempt for fellow earthly creatures. Süskind's murderer had soured on the human race and hated all other people. One must unlearn many things to arrive at happiness, to put an end to the resentment between criminals and detectives. Resentment is a reaction to one's own weakness, making an enemy out of someone else in order to avoid examining oneself. Süskind's character simply could not abide his own scentlessness and sought to hide it. A detective dreams of riches and then is mad at a criminal for acquiring riches by any means necessary. Criminals break the laws made by humanity, not the laws made by nature.

Dualism begins to fade when one recognizes that the will to power remains because every living thing interdepends and communicates. These communications are never inherently dark and so the light/good dark/bad dualistic approach to judging them is not wise. Perhaps Süskind's murderer could have been redeemed by creating a perfume that brings on world peace or eternal life or some other fabulously utopian good, but the reclusive writer who has only given four interviews since the book's publication instead chose for some reason to affirm a world where no character's deeds can go unpunished. The perfumed among us need not be construed as villains by the detective novel or as inevitable lovers by the romance novel, but that is the supposedly satisfying convention.

Fortunately, Nietzsche reminds us that we have ample opportunity to redeem our will to power from the easy seductions of dualistic thinking. The third accord of *Zarathustra*, eternal return, will be most familiar to fans of fragrance as well as genre literature. It simply means that life is full of infinite chances to remain open to the present moment without judging it, without angling to solve the conflicts existing in it. Such is life's mystery, almost unbearable to the perfume industry that operates like a mystery novel. When one perfume succeeds in the marketplace, not only does the company capitalize on that by adding a bunch of flankers with slightly tweaked formulas based on its original, but one hundred other companies will ferret out a rough version of that original formula and go on to create their own imitations in hopes of siphoning some of the profits from the original in an endless series of reboots and knock-offs. Yet consumers in the marketplace do not judge Chanel for engaging in dark communication because the 1984 launch of *Coco* has since then begat *Coco Mademoiselle*, *Mademoiselle Intense*, *Mademoiselle L'Eau Privee*, *Mademoiselle L'Extrait*, *Coco Noir*, and *Noir Extrait*.

Romance and mystery stories are often made into a series. Their use of perfume symbolism becomes constrained by conventions that congeal across repetitions until we expect them as stereotypes and mistake them for moral laws. They repeat endless variations on the same few ideas where just a handful of conflicts constantly recur. This is in part because we feel stupidly satisfied each time by the naïve belief that the resolution of the conflict will last forever. This is a ruse, as more novels will publish more variations of the conflict. Yet we do not judge Louise Penny for engaging in the dark communication of producing sixteen different copies of Armand Gamache to foil sixteen different criminal enterprises. Imagine Sherlock Holmes for once truly not catching the killer. Even his nemesis Moriarty eventually ceases to be elusive. And we never tire of respecting Sherlock in equal measure for his keen sense of smell as well as the drug habit he forms to cope with it. Sherlock and Zarathustra are similarly anti-heroic. Nietzsche posits that they may come down the mountain as many times as needed, the smell of their own struggle somewhat endearingly and rather familiarly fouling up all that fresh air. It is an existential perfume that tells a story beyond good and evil.

3 SPACE

In Tunisia, the world's largest producer of orange blossoms, mainly women pick the flowers off the tree one at a time by hand. In Egypt, the world's second largest producer, mainly men shake the limbs of the bitter orange tree until the blossoms fall off of it. Orange blossoms are one of the most prolific ingredients in perfumery worldwide, yet the average consumer of this flower pays no attention to its origin, still imagining north Africa as one indistinguishable mass of desert, unable to point to Tunisia on a map. Off the eastern coast of Africa, beyond the island of Madagascar, Réunion—formerly known as Bourbon—used to be a major producer of superior vetiver until it became too costly in terms of both land management and human capital required to cover the space.

Vetiver is a plant that smells green and woody or earthy, sometimes clean and other times smoky. An unruly beast at the molecular level, it can provide a sturdy skeleton to boost the longevity and minor notes within other fragrances. Yet it has so many facets that can emerge in such a wild and unpredictable manner that it is said to be a perfume unto itself. It's full of contradictions and is one of very few naturals that has no decent synthetic counterpart. Usually perfumers resort to a variety of things cobbled together into an accord that is said to give an impression of vetiver. Like coffee and champagne, the key to vetiver is its location and the soil in which it is grown.

A grass with sharp edges that grows one meter up into the air and one meter deep into the dirt, vetiver began in southern India but is now cultivated mainly in Java and Haiti. How it came to settle on these islands is a classic tale of French colonialism. The seeds of *Crysopogon zizanioides* are sterile, making it a plant that is easy to control. It only reproduces with the help of human hands to make cuttings: macheting the tall greenery away, pick-axing around the base of a clump of vetiver, pulling the enormous and tangled claws of its root structure out from the sandy and inhospitable soil, and finally plunking the nub of root stock seedling back into its hole to grow anew. It survives twelve months in this dirt, through four seasons that alternate between the hurricanes wreaking havoc on everything else but the vetiver, and extremely dry spells that produce the most fragrant essential oils. "Vetiver" comes from the Tamil words *vetti* meaning "to tear up" and *ver* meaning "root."

To tear up this root is back-breaking labor. In Haiti, the annual production of eighty thousand tons of vetiver employs between thirty and forty thousand people. It takes about twenty-five pairs of hands an average of three weeks to harvest a hectare of vetiver. It then takes about ten hours to turn one ton of vetiver into between five and ten kilograms of essential oil. The market value is $350 per kilogram and only going steadily up, so that many Haitians running small farms think of it as a savings account. The price has risen consistently due to climate change, as hurricane damage to the island makes vetiver production far more costly. Haiti has had vetiver demand cornered since World War II, when Indonesia's supply chain was broken by Japanese occupation. Java's vetiver is said to be of lesser quality anyway, given to the more smoky and leathery facets of the root, as opposed to the earthy and even citrusy notes in Haiti's supply. In Java, they plant the grasses in volcanic ash, which likely accounts for the harsher notes.

Planting it in ash makes the job of tearing up the plant's roots far easier, an efficiency that is a good way to maximize profitability but is a dumb limit on the very thing vetiver does best—prevent soil erosion. Vetiver reaches deep into the ground and holds the soil fiercely. So much dirt clings to the root structure after harvesting that vetiver holds the odd distinction of being a fragrance material necessarily priced by volume. One metric ton of vetiver root can contain anywhere from three to five-hundred kilograms of dirt still stuck in it. The farmers do their best to shake out the dirt and make an advance when the roots are delivered. But the bulk of their profit isn't seen until after distillation, when the manufacturer learns how much root was really in there. These roots, clinging fast to the land despite consistently bad weather and frequently unstable politics, face valuation based on how much fragrant oil can be squeezed out of them.

But locals value the whole plant. The leaves get woven into baskets, mats and other useful items that assist in self-determination and gainful employment. The pulp of leftover leaves is used in making paper. The roots are hung over open windows and spritzed with water to function like an air conditioner. Sachets of the roots work like moth balls and make good insect repellent. Vetiver is a popular milkshake flavor in India, a hot tea in west Africa, and a staple of Ayurvedic medicine. It is burnt as incense and used in voodoo rituals. But none of this is as profitable as perfume, so all of this is just a sidebar nicety to the foreigners. If a non-governmental organization like Vetiver Network International wants to sit nearby and track all the farms for some big study on how to use vetiver as a soil management tool, the perfumers are happy to help fund that because environmental protection is good for the brand.

The insect repellent properties of vetiver mean it can be cultivated entirely without pesticides, so even though only something like three percent of the farms are certified organic, basically one hundred percent of them truly could be. Also good for the brand. The label of natural is another coveted one. There are over fifteen thousand tons of vanillin sold every year and less than one percent of that is truly sourced from natural vanilla. It can cost as much as four thousand dollars to manufacture one kilo of true vanilla, while the equivalent kilo of synthetic vanilla costs just fifteen bucks. Except to manufacture this one kilo produces around 160 kilos of waste and the product itself cannot be marketed as natural. So flavor and fragrance industry scientists ended up splitting the difference between waste reduction and cost reduction by pioneering a biotransformation process where "natural" vanillin can be produced through bacterial fermentation of rice bran, curcumin, or even cloves with less impact on the environment. There is a cornucopia of offenses one might commit against the earth in order to capture the molecules necessary for making perfume, which brands increasingly answer for in a variety of ways: cruelty-free guarantees against animal testing, certification that the fragrance is vegan or all natural or alcohol-free, and all kinds of sustainability undertakings.

For example, the only way to get resin out of a tree is to do minor damage to the tree so that it produces resin to heal itself. Thus, overly aggressive tree tapping must be prevented from sliding into wholesale deforestation. Another example is synthetic musks, which were invented out of necessity when over-exploitation of the musk deer sent its population into an extinction spiral. Many civet and musk naturals are now banned from use in fragrance worldwide to protect these species, and the synthetic versions are built strong enough to survive in the average washing machine, so the benefit of saving a species has been at the cost of all that laundry detergent flushed into rivers. Of course, the process of manufacturing fragrances is itself very far from carbon neutral and there is a lot of debate about the extent to which perfumes themselves are poisonous, full of volatile compounds or making extensive use of petrochemicals that seep in through the skin.

Then there is the idea that perfume is air pollution. Personally, I like perfume that possesses a tremendous blast radius—something with a sillage that takes up an entire meter of airspace around my body and an eternal longevity that still lingers after eight hours in the heat of life. People sometimes know I've recently been in the room. People sometimes inhale my presence before they apprehend me with their eyes or ears. I project my barbaric yawp into the atmosphere of our mutual existence. It is transcendental, this creeping of my molecules into the zone of your molecules, barely consensual and totally inevitable because breathing is autonomic—we breathe without paying attention, we take molecules in and expel other molecules all without thinking. Perfume is an invasive thing. It colonizes the air, politicizes territories of oxygen. It is astral, Olympian, supernatural, Elysian. Perfume is violently awesome.

During the coronavirus pandemic, I was masked up while essential working due to the presence of about a hundred other student bodies spending an hour at a time in my thirty-five by thirty-five-foot windowless cinderblock box of a classroom each day. So I kept wearing fragrances that would be considered ridiculously bold choices, scents that were not safe for work. The office is supposed to be a quiet place where nobody is intruded upon by the loud perfumes of others. It is a shared public space that still clings to some mythic notion of individual olfactory privacy. But there I was, sweaty with the anxiety of stumbling into did-I-just-get-infected moments of paranoia every single day. Sometimes nothing made me feel sane or able to solider on except now and then surrendering myself to the overpowering embrace of Guerlain's *L'Heure Bleue* or Montale's *Full Incense* seeping in through my face mask. I kept to scents that offered syrupy warmth and ancient everlastingness, that supersaturated the air around me with smoke and flowers, a percolation of thick comfort that felt like it might help defend my vulnerable personal space against the invading virus.

The kids liked it, too. If they were going to risk leaving the safety of their quarantine bubble to be in my company for an hour, getting a good whiff of some weird fragrance they'd never get to smell at home was about the most reasonable adventure available. My main daily audience of teenagers has an innate respect for the not-so-polite perfumes, either because their own shelf is limited to Axe and Febreze, or because they're still figuring out how much they like to slap on in the morning. There's a certain sublimity in overdoing it and owning that, like walking into a trendy restaurant in the Eighties doused in *Giorgio Beverly Hills*, weaponizing oneself and launching into the probability of notice by others. It is a youthful and even merciless attitude, a joyous instigation of possible aesthetic conflict, a strategy for swinging away at immortality. And I did use it to establish control over my students, to telegraph them important messages about our classroom vibe for the day: big citrus energy when introducing a new concept, white flowers and fresh greens for getting into the thick of things, leather on deadline day, fruit and amber to smooth any difficult one-on-one conversations.

An administrator once walked into my classroom and paused halfway in to ask whether something was burning. He'd simply caught a whiff of my black pepper top note, but I could see in his expression that he was undergoing some kind of pointed emotional terrorism and the first thing that popped into my mind was how Cleopatra is supposed to have perfumed the sails of her ships. Imagine the effect this would have had on those waiting at the port: the legendary and long-awaited ship emerging gradually over the horizon line, then suddenly waves of myrrh, cinnamon and cardamom so strong they can overtake the smell of the sea air itself. Especially in the context of a first impression of an Empress, this is a brutal and intoxicating display of power, so shocking it would transfix the memory instantly. Even Shakespeare was much later drawn in to writing about it.

Like those who work on boats or who train cadaver dogs, perfumes cause one to consider the minutiae of the terrain. I shake some fragments of orange peel onto baked sweet potatoes and read the label to see the peels come from Valencia, California. Perhaps the city is named for its orange groves and perhaps those groves are full of fruit whose genetics hail from the sister city in Italy. Perhaps I branch out into thinking about all the places oranges are grown or all the types of oranges there are. These are the globalization rabbit holes one goes down when one has an interest in perfume.

Some perfumes are like a tuxedo and other are like a raincoat, and it matters where one is going to wear them. As a result, the perfumed may think about weather, temperature, season, wind conditions or humidity more than other people generally do. Once in my life, I achieved an immaculate vacuum of conditions. I'd lit a joss stick for my afternoon meditation, closed my eyes and began. Ten minutes or so passed by and then I opened my eyes. When I did, a white line ran down the center of my vision. I thought I was hallucinating, maybe seeing some kind of aura. My blinking disrupted it and it was gone. It was the plume from the incense, moving upward in a momentarily perfect column. That's how still I had been and maybe the closest I will ever come to nirvana.

Many spiritual traditions use smudging, especially with sage, as a ritual practice to cleanse a space. Even folks who do not actively invite extra scenting into their homes do at least consider how some scents ought to be avoided. In the same way they do not want to live by an airport, people also do not want to live by a sewage treatment facility or a hog butchering plant.

The geography of any given subculture orchestrates a foundational scent palate for that group's perception, for example, through the prevalence of particular food crops. In Columbia and Brazil, the Tucano are an indigenous people that distinguish between the jobs of various tribal members by the different smells associated with hunters or farmers. In Senegal, the Serer-Ndut people use scent to organize their people into subgroups, and they associate Europeans with the smell of urine thanks to an unsuccessful attempt by French colonizers to wipe them out.

These useful methods of sorting and group formation are often eventually wielded as xenophobic stereotypes, such as the notion that all brown-skinned taxi drivers in New York City smell like curry. Scent marks an immigrant even though, like amber, curry is a heterogeneous mixture of many options. Curry is an ever shifting, seasonal and regional blend of things like saffron, turmeric, fennel, cumin, nutmeg, tamarind, sesame seeds, coriander, chilies and more.

In the West, the primary impulse has been to deodorize. Emphasis is placed on masking bodily odors out of fear of being embarrassing or disgusting, creating a mission where the language is focused on eradication. Imagery surrounding perfume is often borrowed from conquest or warfare, too. There's *Flowerbomb* and *Spicebomb* with their hand grenade shaped bottles, or Jean Patou's even less subtle approach to the smell of *Colony*. The 1938 bottle was shaped somewhere between a grenade and a pineapple, while the leathery scent itself was a mix of pineapple and rubber—boastful about these being two of the new products being exported from French colonies, complete with an ad campaign featuring a pair of dark eyes framed by dark skin peering out from the shadows. In the BBC television series *Killing Eve*, there's a great thread in season three where serial psycho Villanelle prepares for reunion with the detective trailing her by commissioning a fragrance that she imagines should smell like Roman bloodlust. Cosmetics have always been universally understood as warpaint.

Although the United States is the third most populous nation in the world and the second largest market for perfume, the biggest market by far is the sixth most populous nation of Brazil. There are over two hundred million Brazilians and their culture is fanatical about taking showers, to the point of three to five showers per day, after each of which there is liberal application of perfumes. Some of this is just due to heat and humidity of the tropical climate, but there is also a lot of pressure or pride associated with maintaining cleanliness in even the poorest communities. Colognes and deodorants do a brisk business under these conditions, but luxury perfumes only account for a tenth of Brazil's market because of the massive taxes levied on such imports. The average Brazilian owns three bottles of perfume and in Brazil it is still possible to make a living selling fragrance door-to-door, since fully fifty percent of perfumes sold there are moved in this manner.

Beyond a robust appreciation for personal hygiene, Brazil also boasts the lion's share of the Amazon rainforest, which comprises forty percent of the world's remaining rainforests and a correspondingly enormous amount of carbon storage. The Amazon is thus a linchpin of climate change's most catastrophic scenarios and one of the most biodiverse places on earth. Imagine all the uncharted smells in there. That's why Symrise invested over six million dollars to put a research facility right at the edge of the jungle, a forty-hour drive or four-hour flight north from its regional headquarters in São Paolo.

The global fragrance market is worth about fifty billion dollars and Symrise is one of the five big firms that share about sixty percent of the marketplace. It's the only one with headquarters in Germany, founded there in 2003 through the merger of two firms that had been around since 1874. There are two Swiss firms in Geneva that were both founded in 1895, Firmenich and the world leader Givaudan. Takasago was founded in Tokyo in 1920 and International Flavors and Fragrances (IFF) was founded in New York City in 1889. Despite the locations of the top dogs, there is a tendency to associate perfume with France. Indeed, there are still many small and family run manufacturers of natural raw materials that have been located in the south of France since the eighteenth century. This region, in Grasse, employs several thousand people.

It's a nine-hour drive or ninety-minute flight north to Paris, the iconic metropolis from whence all fashionable objects must for some reason emanate. Because it is home not to the manufacturers, but to the brands. Guerlain, certainly, or the thousand ships launched by cosmetic juggernaut L'Oréal: Garnier, Lancôme, Cacharel, Ralph Lauren, Yves Saint Laurent, Giorgio Armani, Atelier Cologne, Viktor & Rolf, Diesel, and Valentino. Those are just the parts of L'Oréal that sell proper perfume. The scent of lipstick—of lotion, of shampoo—also comes from somewhere.

4 TIME

Threw in the towel on my high school chemistry class. There was too much math—plus the widespread assumption that girls lacked aptitude for science—so I slipped through the cracks. This was about a year after my first attempt to mix a perfume at home. Someone gifted me a kit with base alcohol and a few juices to dose out as I saw fit. All I did was dump in the entire vial of strawberry. My teenage investigation began in a minimalist one-note place. Like learning that first chord on the guitar and hitting it a million times in a row, I shook up the bottle and sprayed it all over myself immediately.

Two minutes of experimental joy ultimately begat months of field research as I became a walking strawberry for many moons. I studied this one note from every mental angle, in a variety of weather and scenery, day and night, until all possible means of conceptualizing it were exhausted. From the very beginning, my instinct was that a perfume could fascinate—could capture imagination and withstand scrutiny—nearly endlessly. The scent suddenly appears in one's nostrils and the mind is whisked away to other times. Beneath that sense of infinity is also a shocking immediacy. It's a stealth ninja creeping across the air until it's a thunderbolt right inside one's own head.

The academic time it takes to construct that thunderbolt is enormous: all the math and science classes one must ace to learn the process, then the actual lab work needed to create it. Except that so many of even the most prestigious perfumers report they are self-taught. They are like me: wrestling with my little home kit, flipping through pages of textbooks with advanced equations and molecular diagrams, epiphanies popping up over the years in the form of a few clear paragraphs here and there, where some rudimentary methodologies finally arise from my passionate madness. The discoveries themselves arrive like lightning, the quickness of their revelation hilariously framed by one's plodding pace of study.

After years of self-directed learning, now I could say that my one-note strawberry wonder was a flavorist's notion of strawberry, flatly red and saccharine. Ex-lovers from the time period when I wore it will be struck by a recollection of me when they are at the circus innocently licking a shaved ice, or at home in the kitchen unscrewing a jar of jam to make their kids a sandwich, but not when they go strawberry picking.

A perfumer's notion of strawberry starts with fructone, a synthetic aroma compound that conveys an apple type of fruitiness and is also part of the smell of pineapple. On top of that is ethyl maltol, an organic compound that looks and tastes like sugar. A wilder strawberry sensation would add in methyl anthranilate, found in both grape Kool-Aid and bird repellent. But my basic level of comprehension doesn't necessarily translate into any practical serviceability because I know nothing about how to put these molecules in proportion or mix them together in a way that creates a pleasing smell that can be safely applied to my body.

All these components take time, and a lot of legit chemists with proper graduate degrees to do the grunt work of making the mixes. Even with their superior scientific and mathematical abilities, the chemists are still stumbling around in the dark, hoping to find something profitable. Imagine that the chemical equations for all the available smelly molecules on earth are listed like a giant telephone directory. Back in the days of sequential assignment of land line phone numbers, one could maybe guess at the number of an apartment down the hall if one knew the phone number of some other apartment on the same hall. Chemistry still operates its periodic table in that mostly predictable way, but scent doesn't. So even if one correctly identified all the phone numbers on a given hallway, the smells in those apartments don't run in any sequence. Discovering a new smell is like throwing darts at that phone book. Sometimes one gets lucky and hits a sweet-smelling apartment, but then maybe it can't be incorporated into perfumery anyway because the owner of the apartment is a jerk or the building is abandoned—the new aroma compound doesn't play well with one's other compounds or it contains an ingredient banned from exportation. Sometimes a new smell is accidentally discovered backward; for example, in this apartment we have the explosive known as TNT but add just four carbon atoms to try for a bigger bang and then the tenant in the apartment down the hall is the first synthetic musk scent.

Fortunately, perfumers have digitized the entire telephone directory so that each chemist is not stumbling into the same dead-end apartments. Even after the computer spits out a list of reasonable numbers to call, the chemist must sit there and call them all, hunched over beakers and droppers to precisely count out miniscule amounts of very many ingredients. Perhaps a formula has fifty elements and the lab tech not only doles out all fifty with exactitude, but also all the variations the master perfumer has requested to contemplate, such as a set of ten options where one molecule is increased by a quarter of a percent each time and a second set of five options for each of those ten where the ratio of two other molecules is reduced proportionally alongside the quarter-percent increase of the other. And all the results may smell like garbage.

Despite my awareness of this tedious drudgery, I undertook another single-note experiment during the pandemic in lieu of a New Year's resolution. In need of calming and cleansing vibes, I turned to lavender. Bypassing the prospect of spending hours learning anything about gardening, I bought a small hydroponic system to take the guesswork out of sunshine and water, popped in the lavender pods on the morning of January 1st and waited. They sprouted, grew tall and green, and I took many proud photos of their progress for Instagram. At the end of February, I harvested the tiny crop and bundled each fragrant pile to hang above the fireplace until dried.

Now how to get the lavender into a perfume. There were so many time-based decisions to make. How long to let the plants dry. Instant steam distillation with double boiler method versus the slow and steady sun tea method. Whether I cared for my lavender to infuse itself into the chosen base by the light of a full new moon. Then coronavirus quarantine became paralyzing and I forgot all about it, the perfectly dried lavender still perched in piles hung from the fireplace mantel, my wife and I occasionally stumbling into its cloud of refreshing loveliness. In all the new overwhelmingness of pandemic life, the process of gaming out these choices about how best to use the lavender seemed as stupid as several dear friends' sourdough starter experiments. As they started to admit defeat on the perfect loaf, time began to weigh heavily on me. I'd already invested four months in getting to this point, patiently accomplishing the growing and drying process.

The remaining decisions produced useless anxieties about timing, so I opted to do nothing. The lavender turned to twigs and dust, and that was the end of that, except for a lingering bit of jazziness in the air around the fireplace. What got to me was the prospect that the hard science part would be never-ending. I might ruin the plant matter with bad decisions and thereby snuff out the four months already invested. Or I might make perfectly solid decisions and the result still might not be what I imagined. I could tweak all the parts, each several dozen times and spend the rest of the year experimenting toward a great lavender perfume that may never arrive. Or I could browse down a rabbit hole of lavender interpretations made by other people who'd already spent all that time and sample their efforts.

It did not appeal to me to invest my time in the science part, so I shifted focus to the creativity part and leapt into some mixing experiments that could be done in an hour or two. More advanced than a singular strawberry whiff, but nothing more complicated than a three-chord punk rock song. I needed chords that would go together easily without a bunch of serious calculations, so I turned to the Demeter Fragrance Library. Wife to the king of the gods and mother to the queen of the underworld, Demeter has a high place in Greek mythology. As goddess of the harvest, she is often depicted with fruits and flowers and is a good choice for the name of a perfume manufacturer that is focused on single-note fragrances that express commonplace smells like *Dandelion*, *Espresso* and *Laundromat*.

These notes are referred to as linear because they express an olfactory idea immediately and don't change as they dry down. Yet as I began to put three of these notes together to make little spritzes for some of my friends, they surely did work together in both vertical and horizontal accords to express more complex ideas. Verticals have a typical top-middle-base note configuration, whereas horizontals are built entirely out of note families. For example, even my top note citrus combo of lemon-lime-bergamot proved a three-chord punk song still reveals that perfume itself is like a symphony orchestra. Every perfume is a set of instruments working together, creating varieties of harmony or disharmony, stimulating the senses through variations in tone and volume. Like a symphony, the notes of different instruments come and go. To an untrained nose as to an untrained ear, these instruments sometimes overload into cacophony and the fragrance becomes an incomprehensible noise.

Many of my three or four ingredient Demeter mixes were noisy, but a few were gems, highlighting the odd possibility of hearing a symphony backward—all the notes at once at first, until top and then middle notes gradually fall away into a quieter composition. Smelling a perfume like this is akin to hearing the big finale of a piece of music and then hearing the different sections—woodwinds, brass, strings—drop away one by one. Citrus notes are short-lived and certain flowers go somewhat longer or at least sound out their notes in a slightly more complicated manner. Woods and resins provide a steady and long-lasting structure underneath, exhibiting more than just a homophonic commonality between the idea of base note and bass note.

These changes in composition are not really the work of chemistry; really, it's just time doing its thing. That's a big difference between a symphony and a perfume: one can listen to a recording of a favorite symphony over and over again without depleting it, but perfume lovers necessarily kill their darlings. A symphony never goes bad, although opinions about it may change. By contrast, all perfumes eventually go bad. Maybe it'll take two hundred years to oxidize away within a sealed bottle that is never quite airtight enough to protect the precious liquid, or maybe it'll take a mere twenty-five years with some regular use. Though a perfume has much in common with the structural ideology of a symphony, as an actual art object, a perfume functions more like a cassette tape. If you listen to or spray on your favorite mix a half dozen times a day, you'll eventually wear it out. We do wear out a perfume simply by enjoying it, by letting it out into the open air or burning it away on the surface of our skin. Perhaps this ephemeral quality is what keeps the olfactory arts from being elevated to a level on par with highly respected major forms like painting, poetry and music.

A drying down always points toward the larger decay happening inside our perfume's bottle and the general rules of putrefaction in our natural world at large. Newsflash: things die. Fruits and flowers die; perfumes die; people die. Unless we're examining an ancient sliver of incense wood housed in some museum in Kyoto that probably smells exactly like when it was first collected several centuries ago, but nobody is actually allowed to burn those because they are an incredibly finite resource. Again, even a smell we successfully preserve will someday run out if we continue to attempt smelling it. The corresponding implication is that smells can go extinct.

This is the perennial woe of anyone who sniffed the original formulation of Guerlain's *Mitsouko* or Chanel's *No. 5*.

As the objective compositional ingredients for a fragrance shift due to availability of natural ingredients—the advent of cheaper synthetics, changes in export or safety regulations and so on—the secondary market for vintage fragrances expands. A 75ml bottle of brand-new *Mitsouko* retails for $125. A 50ml vintage bottle from the 1930s, about a decade after the perfume first launched, is worth around $300. A 60ml bottle from the late 1950s with the more iconic shape still used in a streamlined version by Guerlain today is worth around $1,200. If one is just looking for an empty bottle and box from the original 1919 release, that's at least $100.

Chasing smells from the past and finding ways to recreate or preserve them is a hot area of study. Robert Muchembled's *Smells: A Cultural History of Odours in Early Modern Times* devotes forty pages to contemplating the smell of shit during the medieval period. Or there's the Odeuropa Project, a team of scientists attempting to recreate smells of sixteenth century Europe by developing an artificial intelligence that can analyze images and texts in seven different languages in order to build composite portraits of aromatic items. They then fork over these detailed descriptions to chemists, who recreate modern day versions of presently extinct smells to eventually go on encyclopedic display in a museum so everybody can smell the difference between today's Camel Lights and the pipe tobacco of five-hundred years ago.

Recreations of the past such as these, and even the evolution of mass market fragrance formulas over time, beg the question of the ship of Theseus. This is a philosophical thought experiment that asks whether an object should be said to be fundamentally the same object if it has had all of its constituent parts replaced. Theseus was a mythical king and founder of Athens who needed his ship constantly at the ready, causing it to undergo a consistent stream of repairs when not in use. It's a matter of opinion whether this ship can still be considered the same ship if not one original part of it remains. An underlying secondary question is whether an object might have a soul that is somehow separate from its material form. There are four possible resolutions to this inquiry into the nature of objects over time.

The Greek philosopher Heraclitus said nobody can step in a river the same way twice because the water is always replenishing. This means that identity has no fixed position over time and even from within the same bottle, no two wafts of perfume can ever be alike because the air on which the wafts rest always replenishes. Contemporary American metaphysicians like Ted Sider augment this explanation with more current scientific understandings of four-dimensionality and worm holes, arguing a second resolution in that an object can persist through time just as it can exist in space. This means that the character of a perfume one smelled yesterday is just as real and true an identity for that perfume as the character of its smell tomorrow. The perfume is an object that wafts across many slices of time, and this explains why the rarity of a vintage bottle is more expensive than the most current formulation found in stores. One pays extra for the relatively exclusive privilege of returning to a long-gone time slice.

Noam Chomksy takes a different approach to the problem, arguing that it is based on a fundamentally unsupported conceptualization of ships. There is no reason a ship's identity must be psychically continuous, no reason other than our often-dumb human intuition telling us that it is common sense to assume the way things work in our mind is equal to the ways things work in the world external to our mind. In other words, an object has no innate characteristics, but only qualia—only the "what it is like" within our own individual mental state. This means not only that our interpretation of the waft in our nose may have no relation to the object in the bottle, but it also helps explain why two friends leaning over the same smelling strip at a perfume counter each interpret the waft in their own wildly disparate manner.

The fourth resolution to the ship of Theseus relies on common sense, rather than attacking it. English political philosopher Thomas Hobbes thought the ship ultimately becomes a new ship and the latest version full of new parts would be basically unrecognizable to Theseus except for the memory of the original ship still evoked by the latest version. This means that the waft of perfume today can be understood in the context of one's experience with its previous wafts, even if no two wafts are alike. This also explains why a vintage scent hound absolutely never claims that the latest formula is better than the original—because it would be almost disrespectful somehow to say that Theseus would have been a better captain with the newer ship. It would additionally be absurd, because the newer ship cannot exist without the original ship as a predicate. There cannot be a copy of nothing. This explains the disdain many lovers of essential oils have for synthetic fragrances.

But in the case of an extinct smell, synthetics are all we have. Scientific data about wafts that are long gone serves the same function as our individual memory of those fragrances. For example, the American corporation International Flavors and Fragrances Incorporated (IFF) partnered with two artists who have backgrounds in synthetic biology, Sissel Tolaas of Norway and Alexandra Daisy Ginsberg of England, to create an immersive installation called *Resurrecting the Sublime*. They used DNA harvested from specimens stored at Harvard University to recreate the smell of three flowers that went extinct due to colonial activity. The sublimity of this project relies upon the extreme coolness of sniffing the ghost of these flowers in conjunction with the terrifying knowledge that our own irresponsible stewardship of the environment is what caused these smells to be destroyed in the first place.

IFF has always demonstrated an interest in how perfumes expand and contract our notion of time in both the making and the sniffing. In *Speed Smelling*, the annual contest they host, perfumers compete to make quick batches of fragrance with reference to a common theme. Past themes have included "Postmodern" and "Secret Garden." IFF then sells 10ml atomizers of the top ten winners for $185 in a gorgeous box set complete with detailed profiles on each perfumer and their thoughts on the creations. In contemplation of the pandemic, the 2020 contest was titled *Slow Smelling* and its theme was "Luxury, Calm and Pleasure," riffing on Baudelaire's poem "Invitation to the Voyage." The idea was to capitalize on perfumers suddenly all being stuck at home with tons of free time to create, working through the poet's definition of beauty as abundant, calm and voluptuous.

Also inspired by poems from *Les Fleurs du mal*, yet far more reliant on time as an evocation of death than the IFF challenge, there is the work of horticulturist and photographer Roberto Greco. His 2017 series, *Oeilleres*—which translates to "blinders," like the ones put on a horse—looks at flowers in the moment they begin to decay, documenting their gradual putrefaction as an allegory for the equally human condition of mortality. This exhibit was accompanied by a limited edition of five-hundred bottles. A 50ml bottle of *Oeilleres L'Objet Parfumant* is a $215 investment in the sweet stench of death, its generic floralcy eliminating any recognizably lovely notes like roses or jasmine and going instead for a dried and honeyed approach to pollens: eucalyptus, broom, chamomile, lavender, cumin, incense, hay, musk and styrax, according to Luckyscent. According to *Fragrantica*, the middle note involves less incense and more mushroom. Most reviews of it focus on the chamomile notes and classify the perfume as fresh spicy, herbal and green—not quite the aromatic morbidity one expects after viewing the photographs, hinting at the way we love our past when it comingles with our naturally impending doom.

Scent is an archive for memory, yet scent itself cannot really be archived except with great difficulty. If one wants to smell something the way it smelled two hundred years ago, one should go to the many small, local perfumers all over the world who for decades and occasionally centuries have avoided mainstream branding and marketing. These are families that have handed their formulas down generations of bloodline. Their secret recipes can usually be enjoyed cheaply, like those at Bourbon French Parfums. The good people at Bourbon French have been in the French Quarter of New Orleans since August Doussan set up shop there in 1843 and any generous 10ml sample bottle is a mere $9. This means you can try literally everything they have ever made for a grand total of less than $900. I believe over the years I pretty well have. The samples I liked most, I got in 120ml bottles for under $40 each.

This includes the legendary *Eau de Cologne* worn by Napoleon, the incense musk of *Eau de Noir*, and *Olive Blossom*. The Bourbon French interpretation of *Olive Blossom* is how I ended up going to them in the first place. I acquired a scent memory that I became desperate to find in a bottle. While attending graduate school in Baton Rouge, behind the library at Louisiana State University, there was an alcove that was always wonderfully redolent with the sweet olive growing nearby, and I wanted to be able to capture that, to be able to control the frequency with which I could feel all the things I felt inside that scent. It came to stand in for that extremely formative three years of my life. As I combed the internet for renderings of sweet olive, I realized the raw materials could only be sourced from Louisiana for authenticity's sake. There are only so many perfumers in the state, and when I contacted Mary Eleftorea Behlar, I knew she was the one.

Mary, the owner of the shop since 1991, walked me through their options—*Sweet Olive* versus *Olive Blossom* versus *Evermore*—and ultimately sent me all three so I could decide which one most closely resembled my memory in a bottle. Evermore included the *Stephanotis* flower often mistaken for sweet olive. My money was on *Sweet Olive* itself, but it focused on the sugar and wax of the flower, whereas *Olive Blossom* captured not only the flower but the green of the whole plant. That was a good surprise, and I use it like the flux capacitor in *Back to the Future.* Any time I need to reminisce about the person I used to be—or summon her forth to vanquish new challenges for which she is sometimes better equipped than the person I am now—that *Olive Blossom* perfume has proved to be very efficacious medicine worth keeping close at hand for emergencies.

5 TECHNOLOGY

Perfume has always been considered a tool of medicine. The idea of men's cologne sprung from aftershave balm. Before the popularization of full-body daily washing, snake oil tonics arose from a need for antiseptics and scents strong enough to cover the stench of human body odor. Perfumes are still widely used in traditional and plant-based medicine. Since the 1980s, Japanese culture has awarded mainstream merit to forest bathing—walking aimlessly among the trees and soaking up the smells—as a means of supporting one's health and happiness. That type of nature therapy is backed by science, and nowadays we file it easily under the heading of mindfulness practices.

It's a common saying that the best medicine is laughter. Anything that can substantially elevate one's mood for a few moments will immediately reduce stress hormone levels. Perfume can certainly provide this remedy on par with laughter. An interesting or weird or even gross fragrance can even produce laughter—ask any kid who ever farted in another kid's direction. Every technology has the potential for weaponization. One may be skeptical of aromatherapy and classify it as an alternative or fringe belief, but the marketplace begs to differ.

Resorts and retailers spend plenty of money on scenting their spaces because data shows that fragrance makes us do things. It works like up-tempo music, playing softly in the background as one picks through racks of clothes, subtly compelling potential customers toward spending more money. For example, my wife and I went to Vegas to get married by an Elvis impersonator and we stayed at The Venetian. The resort utilizes a cold-air scent diffusion technology hooked up to the ventilation system to spread its signature fragrance throughout the hotel and casino. They have over two million square feet of airspace lightly decorated with a fine, dry mist manufactured and installed by Aroma Retail, which provides fragrance for many properties in Vegas. The Venetian's scent, *Arancia*, is especially well known for its massively Mediterranean dose of orange citrus tempered by some floralcy, with a faint undertow of nutmeg and clove.

The energy of an orange citrus scent uplifts the mood and focuses the mind. Casinos like for their patrons to be alert without tipping over into anxiety, to stay positive despite the house always winning. *Arancia* is fun, outgoing and invigorating. Aroma Retail also does a brisk business in home fragrance, so if one wants to bring The Venetian vibes home, all it takes is $200 for the machine and the first 120ml bottle of fragrance is free. After that, a month's worth of *Arancia* is $40. Aroma Retail also sells 120ml room sprays in a five-pack for $60 and I'm thinking about ordering it for our upcoming wedding anniversary, which we will need to find creative ways to celebrate during the ongoing coronavirus pandemic lockdown.

That would nicely set our scene, but ultimately, *Arancia* is a signature that doesn't belong to us. Its medicine is meant to work in public much the same way as voodoo's concept of private mojo, which could include things like a personal charm or gris-gris bag. This moves me to contemplate what medicinal impact a perfume centered on the story of my marriage might accomplish, how it could reflect our two individual selves and continually call us to remain bound together in all that we do. How one perfume could be a technology for both of us that serves to promote the steady renewal of our vows and operate as a reminder of mutual affiliation, a secret wedding band winking at the evolution of our sacred bond.

As the more extroverted of the two of us, Mindy's personality should steer the top note. Her big smile is the one we use to greet the world. She is an orange blossom, the flower on an orange tree that is intelligible to most cultures as a sign of good fortune. The tree went from China to Spain to Florida, where orange blossoms became the state flower. As a bonus side-story for a marriage-focused perfume, orange blossom is still commonly associated with weddings. A beautiful irony of the flower is that its tree produces bitter oranges that are not particularly edible, so even though one looks at this bright tree and wants to gather its fruit, the more approachable gem is its flower. It requires warmth to grow properly and smells sweet yet still fresh. Fragrance is made of this flower in one of two ways with quite different results in the smell, highlighting the extent to which technology manipulates even the natural components of a perfume.

One method of getting fragrance from orange blossoms is solvent extraction. The flower petals are washed and macerated in a solvent like hexane. This alcohol strips the flower of its smelly molecules along with some waxy substances and plant matter, resulting in a concrete. This lump is then washed with another solvent like ethyl alcohol until only the highly concentrated and deeply fragrant oil is left behind, which is called orange blossom absolute. A few tiny drops of absolute go into the perfume bottle along with some water and a lot of body-safe alcohol. It takes one thousand kilograms of flower to make just one kilogram of orange blossom absolute. One milliliter of this pure juice is worth between $20 and $30.

The second method, which is somewhat less expensive, is to distill orange blossoms in boiling water and collect the fragrant steam. In the pool of condensation, the flower's essence floats to the top of the remaining water, and this collected oil is called neroli rather than orange blossom. The distillation method results in a much greener and therefore fresher scent than the solvent extraction method, and one kilogram of neroli oil can be had out of just 850 kilograms of the flower.

This begs the Lévi-Straussian question of cooked versus raw, asking how technology adulterates our experience of a perfume's natural ingredients. Just as varying body temperature causes a perfume to behave differently across a diversity of skins, varying intensity of heat application causes the flower's essence to behave differently across a diversity of extraction methods. Distillation by water is a less harsh and more natural approach than scrubbing a flower with alcohols, but it also occurs at a higher temperature. If a perfume's authenticity centers on how close it comes to the experience of sticking one's nose in the flower, orange blossom absolute is more authentic than neroli oil. On the other hand, neroli oil may be closer to the experience of getting a whiff of the entire tree.

One generally thinks of essential oils as quite natural because of the purity of their single ingredient, but even essential oils are not raw. Technology is always already in operation in the cornucopia of highly sophisticated and precisely calibrated techniques for pulling fragrances out of the world and putting them into a bottle. A perfume is, by definition, a cooked object. There is a necessarily civilized component to it, even in just thinking about the connection between perfume and fire—the best-smelling cave-dweller was surely the one who tended to whatever was roasting on the prehistoric barbeque.

In our house, I do the cooking. My wife asks everyone to the party and then my job is party logistics. She is the thrilling invitation and I am the prime mover, the heart note that puts meat on the bones of the general structural direction we've chosen. As much as she has always locked in on sparkling, exuberant perfumes like *Orange Blossom* by Jo Malone, *Fleur d'Oranger 27* by Le Labo and *Fleurs d'Oranger* by Serge Lutens, I have a serious attachment to anything that smells smoky. Perfumes that foreground the technique of smoking have a very wide range, from burning rubber tires to cigarettes to maple campfires to incense. My house is full of candles and incense, empty of room sprays and plug-in air fresheners.

The heart note for this marriage perfume I'm dreaming up takes the form of a smoldering resin. All conifers produce resins: cedar, fir, pine, cypress, juniper, redwood, spruce and very many more. This is thicker, stickier, and compositionally different from sap, the sugary food that all trees make through photosynthesis in order to grow. Resin is a tree's medicine for itself, made to protect against pests or pathogens, promote healing within the wood when it gets injured, and excrete mineral waste to cleanse the tree. As such, resin is an exemplar of self-care and self-sufficiency that at the same time thwarts any easy classification. It is semi-solid, semi-translucent, semi-soluble: benzoin, camphor, copal, dragon's blood, elemi, galbanum, labdanum, opopanax, turpentine, Balsam of Peru, Balm of Gilead and very many more.

Of all the things one can light on fire in this world, perhaps there is nothing so appealing as frankincense, widely considered the original gangster of perfumery. Frankincense is a resin that comes from five different species of Boswellia trees that are found in North Africa, the Arabian Peninsula, and India. The tree is so hardy in adverse desert conditions that it sometimes grows out of solid rock. It can get as tall as twenty-five feet and after about a decade, it begins producing resin that can be tapped a few times per year. Each year, the Roman Catholic Church buys about fifty tons of frankincense from Somalia. Although frankincense has been traded for over six thousand years, it went viral in Western cultures after making Biblical headlines as one of three gifts given to Jesus when he was born.

According to the New Testament, Eastern scholars gave gold in homage to the virtuous kingship of Jesus upon earth. They gave myrrh, which was commonly used as anointing and embalming oil, in acknowledgement of the suffering and death that he would endure. And they offered frankincense as a prayer to the deity of which Jesus was a part. By the time of these gifts, every religion on earth had already been employing frankincense as an offering of spiritual medicine. This was in addition to its ongoing widespread use as an anti-inflammatory in traditional Chinese medicine and Indian Ayurveda. The most current studies are looking into it as an anti-cancer agent. Modern research says frankincense is somewhat useful in the treatment of rheumatism and colitis—two conditions I have.

A heart note of incense therefore speaks to my generally thinky nature as well as the specific kinds of inflammation to which I'm prone, just as a top note of orange blossom speaks to my wife's boisterous nature and her hyperactive attention deficit. The base note of our marriage perfume is perhaps the trickiest because it must reflect how our connection works, the total arc of our growth across a lifetime together. I wonder what John and Yoko would do, or Beyoncé and Jay-Z. Those are the productively codependent partnerships that inspire us: two strong individuals whose clearly mutual influence is so optimal that the marriage itself comes into view as evidence for a conviction that one lives one's best life by fiercely loving another.

As a medicine, the marriage smooths everything and sustains anything, making something everlastingly mythic out of the nothingness of two mere mortals. I think our base note is Iso E Super, an entirely synthetic molecule. Synthetic smells are dreamt up in labs, rather than culled from nature. They are often found by accident. A chemist searching for a musk is likely to end up making a note with a woody or amber character instead. The first synthetic fragrance molecule, vanillin, was made by German chemists in 1874. They immediately formed a company centered on selling it, and Symrise retains its top five status in today's flavor and fragrance market.

It was not until 1882 that a synthetic molecule appeared in a perfume. This pioneering honor belongs to the French. Paul Parquet, a joint owner of Houbigant and nose behind many of its early creations, changed perfume history forever by launching a perfume that not only was first to use a synthetic smell but also first to break away from the concept that a perfume must smell like something in nature. *Fougère*

Royale was a mix of lavender and oakmoss, and it employed the vanilla-ish, hay-like scent of artificial coumarin, which occurs naturally in tonka beans and sweet grasses. The perfume was built to evoke a fern—ironically, a plant that has no smell. Parquet's break away hit spawned its very own fragrance family, the fougères, which are still the bedrock of men's fragrance lines. Fast forward a century and its preeminence is apparent through its most iconic descendent, *Drakkar Noir* by Guy Laroche.

Synthetics add another layer to the question of authenticity because they can be evocative in ways that have no proper origin point in nature. With an essential oil, one might evaluate how much the derivation smells like its plant matter source. Yet one is still compelled to ask this question of smells that have no ancestor. Some part of the medicinal power of perfume stems from how it conjures a material or momentary antecedent. One always wants to ask what something smells like, so a synthetic perfume built around dirt leads easily back to beets for comparison, but artificial grape or banana is not at all like actual grapes or bananas. A technological approach to perfume requires one to consider the philosophical value of similes.

Iso E Super smells like a lot of different things to different people, often somewhere around sandalwood or cedarwood territory. For me, it has almost a sawdust quality, dry and diffuse, more a feeling of volume or density than a smell. It is beige and sheer, velvety almost like pantyhose rendered in a fragrance. It stays close to the skin, releasing its scent very slowly, sometimes noticeably and sometimes not, kaleidoscopic in its hints of pepper or violet. It's also strongly fixative of other odors, helping them to last longer and burn more brightly, growing the presence of other molecules rather than competing with them. The neutrality of Iso E Super enables an excellence in the things it envelopes, confident in its own light but lasting touch, its understatement as a team player.

Two American chemists filed a patent for it in 1973. When it expired twenty years later, all fragrancers gained the right to make it. As such, Iso E Super can be found in small or large quantities in very many perfumes: the industry now produces an estimated 1,800 tons of it per year. The most shocking perfume made with Iso E Super debuted in 2006, just a few months after we got married: German perfumer Geza Schon launched his Escentric Molecules brand, each bottle highlighting a singular molecule. His *Molecule 01* perfume contains 100% Iso E Super—radical and moderate at once, the ultimate minimalist statement that nevertheless generates infinite awe, constantly transforming but never getting any less mysterious. One has to wear it, to experience it firsthand, in order to fathom it at all. If that's not a simile for the ongoing success of my marriage, I don't know what is.

Any difference between medicine and perfume was not established until 1810, when Napoleon declared that apothecaries had to reveal their formulas and perfumers did not. Today, the entire fragrance industry is worth around

$15 billion—whereas a coronavirus vaccine alone is expected to rake in a net profit above $40 billion. Even fragrance applications for already profitable types of machines are highly unlikely to reach such revenue. For example, the kid who won the 2014 Google Science Fair with his Sensorwake invention—an alarm clock that includes a smell component with relaxing fragrances designed to promote gentler waking up—raised $250,000 and got retail space in Target stores but never achieved market traction or even widespread audience amusement. Same for the guy who invented the Wake N Bacon in 2011 and asked *Shark Tank* for a $400,000 investment deal that was not sealed. Same for the epic battle between AromaRama and Smell-O-Vision to launch fragranced films in 1959.

There seems to be a consensus that these other objects, whose dominant sensory focus has always been eyes and ears, cannot be improved by an added consideration for the nose, that these technologies are maxed out in their present sensory forms. Perfume is not perceived as a useful function in these others forms of technology. As the saying goes, form should follow function. But of course, perfume too has functions, so one should consider the form a perfume ought to take. The chemistry is one such form fitted to perfume's medicinal functions, but there is also the artistic function of perfume—the more creative function for which a perfume's bottling has the big responsibility of being evocative.

The bottle is its own work of art, separate from yet cooperative with the treasure inside of it. I inherently mistrust a fragrance if it's in a container that has no creativity or message of its own. The experience of a perfume must be total. Otherwise, it's like going to the best concert ever but sitting in folding chairs. Or it's like amazing haute cuisine but served on paper plates. Ask Steve Jobs how important the sleek, minimalist look of an iPhone is in relation to its functionality and ultimately its market share. Just because a perfume bottle has utilitarian purpose doesn't mean it shouldn't also have aesthetic value worthy of contemplation.

Some people are so keen to consider the bottles that they don't even care about the juice inside. A membership in the International Perfume Bottler's Association costs $55 annually for those collectors, dealers, historians, appraisers, and other specialists who focus only on the fine art of the bottle. It includes a subscription to *Perfume Bottler's Quarterly*, entry to the world's largest annual convention and auction devoted to perfume bottles, a membership directory, a lending library, and a monthly newsletter.

I myself have been pulled into a purchase by the bottle alone: *Bang* by Marc Jacobs. It wasn't even a three-dimensional hustle. I saw an ad announcing the fragrance launch and couldn't tear my eyes away from its caved in, pleated silver structure. It looked like a sheet of aluminum foil in mid-crumple, so shiny and opaque that it would give off a distorted reflection of one's face rather than reveal the liquid inside. The bottle looked like it had been banged—major vibes of gun powder and attitude, only marginally attesting to the scent housed within. The perfume didn't prove to have a metallic accord although there was a good deal of pepper in the top note, which was heralded as innovative and daring at the time much like the bottle was.

Other times I have come to appreciate the role of a bottle only belatedly after loving the juice. Elsa Peretti is an Italian jewelry designer who made the bottles for fashion designer Roy Halston's *Z-14* and *I-12* in 1974. She was famous for an unusual adherence to organic, natural forms. On the shelf, these perfume bottles look like blobs of glass in neutral shades of green and brown, nothing stunning or special like *Bang*. But the large, awkwardly curling square of *Bang* is deeply impractical by contrast, difficult to hold and sometimes requiring two hands to spray at certain angles. One has to pick up Peretti's bottles to understand: they quietly conform to the shape of one's hand and are so easy to spray every which way, settling on a highly functional, even comforting and nonjudgmental form. A bottle of *I-12* or *Z-14* doesn't care whether one is right or left-handed, doesn't care about being gendered, doesn't need to show off. It lets the perfume do most of the talking. The bottles are as effortlessly confident as Halston himself and the perfumes he deployed to extend his market dominance.

6 PERFORMANCE

In the spring of 1994, I was an eighth grader glued to the television as tributes to dearly departed Kurt Cobain began rolling out. The messiah had gone, but his message of grunge continued to proliferate everywhere, and for a few years it was possible for young women to wear Doc Martens and flannel shirts without everybody presuming they were lesbians. We dressed like dudes and everybody wore the same perfume: *CK One*, barely fragrant but massively ideological. The ad campaign was starkly minimalist, black and white with almost no sound, just a motley crew of long-haired people in roughed up jeans and unbuttoned flannels quietly slouching around and looking semi-confrontationally at the camera. Through this marketing my thirteen-year-old brain instantly processed the foundational postmodern idea that there was no difference between a man and a woman.

CK One was the first fragrance explicitly marketed as unisex. Calvin Klein went to Firmenich and asked the firm to make something that could be worn by anyone. Alberto Morillas and Harry Fremont proceeded to craft an aromatic citrus fragrance that defined the lowest common denominator of perfume: basically, a vodka tonic with lemon twist.

At its peak, *CK One* was turning a profit of more than $90 million annually. Despite its very light touch, the original contained more than twenty notes that could be pushed and pulled in so many directions that Morillas and Fremont were ultimately able to spawn about a dozen flankers. These were each branded in a way that carried forward the missions of grunge, even though explicit associations with Nineties counterculture were quickly dialed down: *CK All*, *CK Everyone*, *CK One Graffiti*, *CK One Summer*, *CK One Chinese New Year*, and many special collector's edition bottles.

The grunge association ultimately had to go because Calvin Klein realized he'd launched the world's easiest office scent. *CK One* has rather sadly and even ironically ended up as the official cologne of unobtrusive corporate employees everywhere. Telegraphing an aura of team player, it whispers proof that one takes care with one's presentation of self but that the presentation shall not be complicatedly fussy or generally loud, and it is most definitely not sexy. It understands what is appropriate for a workplace environment. Human Resources departments approve of *CK One*, which is about as anti-grunge as it gets. But there I sat: on the cusp of puberty, imagining how my girlhood might dissolve before my very nose into this total gender revolution that at least for a moment *CK One* truly was. I never bought a bottle; the ideas in the ad campaign were enough to activate me.

This is in marked contrast to the strict male-female division I saw everywhere else in life, but especially in the marketing of perfume. Davidoff's *Cool Water* had been around since 1988 and was the Axe body spray of its time among the boys at school. Those ads featured tan, ripped athletes with brown hair and board shorts emerging from bright waves, wet demigods presented in slow-motion. Despite the overwhelming popularity of the men's fragrance, Davidoff waited until 1996 to introduce the women's version, which was more fruity than aquatic. I remember loving the original version but being afraid to ask my mom to buy me a men's fragrance, lest she be alerted to the fact that I was a queer. Someone gifted her a bottle of *Cool Water Woman* at some point. She hated it and after the bottle grew a fine layer of dust, I nonchalantly asked her if I could have it. *Cool Water Woman* was socially permissible, a tomboy compromise I made to inch just a little closer to the bottle I truly wanted and the socially unacceptable self it was meant to express.

Sometimes I dreamed of olfactory privacy, wishing to walk around inside a bubble wearing the men's version of *Cool Water* but unable to be smelled by others. I wanted to be stealth, to be allowed to pass as reasonably feminine, to silo off the masculine energies of the scents that most genuinely drew me, to fly my freak flags for my own enjoyment unperceived by a policing public. The closet is not a happy place and eventually I came out of it. My first girlfriend wore *Happy*, created for Clinique by one of the perfumers who would later work on *CK Free*. *Happy* launched in 1997 but was predicated upon some studies that International Flavors and Fragrances did in the Eighties. IFF was building a giant map of moods, testing how consumer feelings could be reliably influenced by certain scent notes. The resulting ScentEmotions database was used to construct *Happy*, and blind testing two decades later continues to show that Clinique indeed found a recipe that makes people happier. It's quite like that song by The Searchers, Madame Rue selling *Love Potion Number Nine* down on Thirty-Fourth and Vine.

Of course, that potion was meant to be swallowed rather than sniffed, but the lyrics clearly state that it smelled like turpentine. A lot of people actually like the piney, resinous smell of turpentine. Belgian "perfumance" artist Peter de Cupere once included a heavy dose of turpentine for an exhibit that reconstructed the scent of Leonardo da Vinci's studio. De Cupere's entire body of work recenters smell in the arts. He has done a series of scratch and sniff one-color paintings with obviously evocative names like "Horse Standing in the Stable" and "Still Life with Lemons," along with more intriguingly mysterious or abstract ones like "Madonna with Child" and "Self Portrait of the Artist in the Evening." During the Covid-19 pandemic, he made "Code Blue 19," a short film examining respiratory distress and deep cleaning where he acts as a sad clown behind a face mask made of shaving cream.

One can enjoy the meaning produced by these exhibits and simultaneously have bad feelings while experiencing the specific smells they offer. As an object, perfume always holds the potential for maximally concurrent pleasure and pain. This is what French psychoanalyst Jacques Lacan calls *jouissance*, the kind of flashy turn-on that is both excessive and transgressive. *Jouissance* is the thrill territory where it becomes difficult to distinguish between death and an orgasm. One becomes overjoyed at losing one's footing, at suddenly tilting one's previously secure sense of cultural expectation. One can hardly help but give in to it, to the dark mysticism that is an experience taking place at one's limit. A truly beautiful perfume blows one's mind.

There's an air of sadomasochism about Big Scent Energy, probably best exemplified by the women's perfumes of the 1980s. I love all these sickeningly indomitable narcotics: Christian Dior's *Poison*, Yves Saint Laurent's *Opium*, Calvin Klein's *Obsession*, Giorgio's *Giorgio Beverly Hills*, and so on. These names themselves testify to a dangerously addictive potential for overwhelm held within the bottle. The ad campaigns all featured stereotypical man-eaters, of course— women wrapped in purple and red, untouchably seductive, possessed of some secret knowledge, sirens with their silent scent always ready to trap. The *Poison* bottle is meant to look like an apple, evoking Eve's temptation of Adam in the Garden of Eden. Its most successful flanker is *Hypnotic Poison*, which spawned several flankers of its own—*Elixir*, *Collector Rubis*, *Diable Rouge*, *Eau Secrete*, and *Eau Sensuelle*. Across four decades, the ads continue to feature naked women draped with snakes, women in billowing purple evening gowns under barren trees, their blood red nails always foregrounding an offer of the bottle to the intoxicatedly greedy eyes of the viewer.

Olfactory artist Clara Ursitti once used *Poison* to stage an intervention at somebody else's gallery. This was 2013, nearly thirty years after the perfume's debut. She sprayed it on twenty-five female volunteers who then attended an exhibition opening. The intervention was unannounced. Most of the women she chose were over sixty years of age. These old broads, who mostly move through public spaces ignored, were suddenly the center of attention. It was impossible to ignore the giant fruity amber floral cloud settling over a space where it was not expected. Surely it changed the conversations that evening, perhaps weakened buyers into spending more cash on the artworks for sale than they otherwise would have, perhaps pushed business cards into unusual hands or pushed different strangers into restroom stalls for a spontaneous quickie.

A perfume is always a performance and is never totally under control, as much as a marketing campaign wants to pin it down precisely. An interesting advertisement will definitely prompt me to investigate a fragrance further, but it rarely does the job alone. Sometimes I browse through websites of perfume houses with the intent to purchase, treating them like a store where the goal of the experience is to open my wallet. Other times I window shop or treat the experience of browsing around perfumes like a museum. I can go down the rabbit hole of *Fragrantica*, a kind of *Wikipedia* for fragrance lovers, for days and days. The trouble is, when I go to a site that I want to treat like a museum I still often end up purchasing because that is the only way to fully experience the object. One has to go beyond the marketing copy, beyond secondary source materials, and really sniff the thing. One has to get into the space, the actual airspace of it. Can't fax a perfume. There should be museums and art galleries devoted to fragrance, places that encourage low-cost study and low-risk experiment, as well as more archival scent libraries like the one and only Osmothèque in Versailles.

Because good training materials about perfume are not cheap and not widely available—and more often found in Japanese or French without English translation—one must too often approach the discipline by self-taught analogies. I learned my fragrance vocabulary from other opinion work in food and music, which was tough. But blessedly, the language of fragrance now infuses my thinking when I go back into those worlds, so that the pairing has become a two-way street allowing me more expansive conversation in each discipline. Artist Maki Ueda would agree, having transformed her formidable kitchen skills into a variety of workshops where foods become perfume and others where perfume becomes food. She has also made many installations that use games theory to engage with the scents of everyday life. For example, she created a labyrinth of scented pendulums that hang from the ceiling in neat little rows, which one navigates through by tracking a singular odor. She then later built a version with wood walls, which one navigates by sniffing out the cedarwood, with dead ends at patchouli, olibanum, and labdanum.

Ueda made another puzzle where one sniffs out a cherry blossom, and another where one operates like a dog, either stamping one's paws on the ground while wearing fragrant slippers that invisibly mark one's trail or else running one's nose along the floor to sniff out these trails of pine, orange, rosemary, et cetera. She created an outdoor installation with five kinds of bubbles that, when popped, emit one component of the scent of a rose. One can examine each of the five facets in turn to deconstruct the flower or stand at the place where they all meet for a complete rose experience.

Many of her exhibits involve sensory deprivation, dark chambers devoid of anything to do except use one's nose as the parasympathetic nervous system becomes increasingly active. In another exhibit, *The Juice of War*, she built a glass bulb that fits over one's head and emits the smell of burned and rotting bodies from the atomic bombs dropped on Hiroshima and Nagasaki. And then there's *Perfect Japanese Woman*, an installation set up simply like a perfume shop with four symbolic fragrances in it—representing motherhood, a woven floor mat, the kitchen, and soap.

The last time my niece was in town, I gave her some pretty expensive perfume samples. She had been limiting herself to a cheap rose oil and was almost embarrassed by the powerful aura seeping out from the nearly airtight vials I put into her hand. She asked how best to use them, said she struggles with accidentally overdoing it. I pushed her to say how she can tell if she is overdoing it, to weigh what the consequences of overdoing it are. Flummoxed, she began to come around to the viewpoint that such an inhibited, constrained approach to her body's capacity to be fragrant was nothing more than her own internalized misogyny talking. She had seen too many perfume advertisements depicting kinds of women she could not imagine herself to ever be. There are all kinds of closets—throwing open the doors of any of them requires a certain amount of nerve. I encouraged her to think about some situations in her life where she would like to feel more powerful, more potent. All women who love perfume learn to embrace the facets of their identity that are for better and for worse characterized as witchy.

But let's keep talking about the Eighties, because as much as it's difficult to think through the contemporary ways perfume continues to play a role in the social construction and policing of gender normativity, the further back one goes, the more amusingly clear-cut these issues become. *Weird Science* will always be one of my favorite John Hughes movies thanks to a quick scene in the middle that is seldom remembered in the grand scheme of the kooky science-fiction plot where two teenage nerds create the woman of their dreams by feeding magazine pages into a computer. Out pops Lisa, played by Kelly LeBrock, one of the Ford Modeling Agency's most in-demand commodities in 1985. A few years later, she did a television commercial for *Brut* by Fabergé, the Sixties cologne whose slogan proclaims it to be the essence of man.

In the movie, the boys want to show off their hot new girlfriend, so they take her to the mall. My favorite scene is where Lisa has gone off to shop while the two boys go to a department store perfume counter. They sample a bunch, flirt with Susan the salesgirl and then buy a bottle of whatever for Lisa. The perfume itself is anonymous. They have the fragrance vocabulary of teenage boys, so they just say it smells nice and that they like it. They tell the salesgirl to set herself up with one of these bottles also, as if any remotely attractive girl can wear just any perfume. Yet the moment simultaneously conveys that perfume is extremely cool, quite fancy. It is part of a mature, popularly elite scene and the boys aspire to be part of it. They imagined, invented and then made real a woman who would wear perfume. Kelly LeBrock is iconic in this role: a sexpot, intelligent, articulate, rock and roll, magic and machine-made. Perfume mirrors that and the film uses it as a way of paying tribute to her. The perfumed are fierce and fabulous people. The boys are dumb and bumbling by comparison, even breaking the nozzle off one of the tester bottles at the store. Lisa is ideal, their perfumed queen.

Did they feed a perfume sample from a magazine into the computer that made her, right alongside photos of Albert Einstein and David Lee Roth? The first brand to include samples in a fashion magazine was Giorgio, spending six million dollars on a scratch and sniff campaign in forty different magazines in 1981, such that noses now commonly associate the perfume with the scent of magazines themselves. Its bold, aggressively sunny, tackily luxurious floral mishmash of gardenia, tuberose, rose, and jasmine has become the scent marker for trendiness. Once it is in one's nose, there it sits for an alarmingly long time. One can and will detect it a mile away. Restaurants began to ban it, putting signs out front that begged wannabe starlets not to ruin everyone's meal by wearing such a pervasive, tenacious perfume. Ladies, please make yourselves smaller to accommodate our other guests, those men who pick up the check.

Another big thing in 1981 was the first celebrity fragrance, *Sophia*, by Coty for Sophia Loren. The television commercial claimed it was the most female fragrance one will ever experience, whatever the hell that means. It was an "oriental" floral full of rose and jasmine, but the dry down was a not-particularly-feminine blend of musk and incense. The incoherence of advertising campaigns like this immediately calls to mind French theorist Monique Wittig's *L'Opopanax*, named for a perfume ingredient and herbal medicine. Her book is like the philosophical love child of Sylvia Plath's memoirish novel *The Bell Jar*—which had been published the year before, in 1963—and James Joyce's *Portrait of the Artist as a Young Man*. Wittig's book might well be subtitled *Portrait of the Artist as a Young Woman*, as it concerns the usual formative experiences of a girl's early education and first romance conveyed in an avant-garde style. Each chapter is its own paragraph, running on and on, often with improperly absent punctuation or jarring switches in narrative point of view. In a poetic cut-up style that paved the way for Kathy Acker, it also contains regurgitations of a bunch of advertising messages that first enable the girl to discover her identity, then find her feminism, and ultimately shape her queerness.

These advertising fragments are meant to help girls find their way in the world, and they surely do, even when one revolts against the messaging of the ads by either a countercultural performance or an ironic engagement in the normative life suggested. Perfume is a functional artwork, like food and clothes, which must beguile the public in order to survive through proliferation of the object. It differs from a work of visual art, graffiti sitting up there on the wall with a ready ability to disrupt whenever one happens upon it, free of any imperative to please its audience. An art object that need not entice is one that is free to provoke. In 1973, Revlon was secure enough after almost forty years of fragrance profit that it dared to launch *Charlie*, a lovely green floral aldehyde that was the first perfume ad campaign to feature a woman wearing pants. But also high heels, of course. One can be both pretty and independent, it seems.

We can even go back to 1952, when Estée Lauder decided to market *Youth Dew* as a bath oil instead of a perfume so that women could buy it for themselves. This was an era where men were supposed to buy perfume for their women. If one had no man, I suppose one had to choose between no perfume on the one hand or a ruined reputation on the other. When Andy Warhol died in the hospital after routine gallbladder surgery, one of the things the hospital returned to his estate was a small bottle of *Youth Dew*. Warhol had an extensive collection of smells, especially women's perfume and particularly bottles from Estée Lauder and Lalique. He evangelized the merits of Estée Lauder's *Beautiful* so often and so widely that photographer Paige Powell dropped a bottle of it into his grave as his casket was lowered in, and Estée Lauder later launched anniversary bottles of it featuring his artwork. This fact is probably the only evidence one needs to cement the links between perfume, gender and marketing. If Warhol was deeply into perfume, then perfume is definitionally part of queer pop culture.

The advertising industry has long been dominated by male executives, but so too has the perfume industry. Coco Chanel may be an iconoclast and an institution in her own right, but she did not possess the technical skills required to give birth to the juice of *Chanel No. 5*—that honor belongs to chemist Ernest Beaux, though the results were not his intellectual property and still would not be considered as such under international law today. Even though perfume is increasingly recognized as a form of art, perfumers are seldom recognized or protected as artists. Their names fade away into history even when their greatest creations remain highly profitable. Consider the case of Germaine Cellier. I would love nothing more than to read an entire book about her life and work, and if my French language skills were even remotely passable, I'd seriously consider such a biographical undertaking myself. Here is what little I know, by way of poorly done fragmentary translations and direct experience of some of her gorgeous creations.

Cellier was herself gorgeous. She wore Balmain skirt suits over blonde hair and blue eyes, but mainly she wore her wit. Cellier was so independent that she was often considered difficult, and when she began working for Roure, her rebellious creativity so consistently annoyed the master perfumer there that they took the unprecedented step of giving her a workshop space of her own so as not to lose either of their geniuses. She had three wire-haired dachshunds—Cleopatre, Felix, and Valentin—and while she lived in a civil partnership with left-handed tennis champ Christian Broussas for thirty years until her death, she was often presumed to be a lesbian by fans and foes alike. There are about five decent photographs of Cellier and she is looking directly at the camera in all but one of them, a fiery suffering in her eyes and a not quite lascivious smirk on her dark red lips. These black and white portraits of the loud but mysterious perfumer are truly the best advertisement for her pointy, pioneering perfumes.

Throughout the Forties, she created scents that conveyed two things about her point of view: that one must dare to overdo it on occasion whether out of libertinage or desperation, and that a life lived fully is about the tension of opposites. The twin fragrances she made for Robert Piguet, *Bandit* in 1944 and *Fracas* in 1948, remain modern marvels. Cellier explicitly dedicated the green leather *Bandit* to butch dykes everywhere and said the buttery tuberose *Fracas* was its femme counterpart. *Bandit* contained an overdose of one percent isobutyl quinoline, derived from coal tar with an intensely sharp rubber-leather-tobacco smell. Cellier herself smoked like a chimney, and this perfume is the perfect complement to boost a cigarette and then fade the smoke away into a soapy, cooling chypre base of vetiver and oakmoss. Between the two queer Piguet scents she did *Vent Vert* for Balmain in 1947, with a stunning eight percent overdose of galbanum. No one had ever thought to make verdant fragrances for women, and Cellier had a monopoly on it for quite a while, continuing her love affair with bitter greenery in a strong thyme-anise-violet accord in *Fleeting Moment* for Balenciaga in 1949 and softening the themes of *Bandit* into *Jolie Madame* for Balmain in 1953.

All her formulas tended to be short, as blunt yet colorful as she was herself, full of dissonance and dissent. Her creations and her personality made her a superstar with the magnetism to attract many famous friends, especially other artists. This is not documented anywhere, but it's a pretty safe bet that she and Marcel Duchamp knew each other, and she was likely inspired by his 1921 assisted readymade project. *Belle Haleine, Eau de Voilette* removed the labels from an original Parfums Rigaud bottle and swapped some letters to turn "violet water" into "veil water," a cheeky nod to the performative capacity of perfume. Duchamp also put his own face on the bottle—appearing in drag for the first time, though he'd used the pseudonym Rrose Selavy before. In 2009, this piece set a record for the highest price of a Duchamp work ever sold at auction, costing eleven and a half million dollars and unseating the famous *Fountain* urinal.

The previous owner was none other than Yves Saint Laurent, whose partner asked Christie's to auction 733 art objects from their private collection after the fashion designer passed away from brain cancer. The auction broke many records and brought in almost half a billion dollars, most of which was donated to AIDS research.

Sometime in the Nineties, RuPaul Charles made a television commercial spoof that operated in the same area of inquiry as Duchamp. In a blonde wig, black mini-skirt, and thigh-high stockings with a face beat for the gods as usual, Ru struts down to the gritty-looking corner in blue-washed footage to herald a new perfume—*WHORE*, "a perfume for who she is." It's pitch perfect and utterly camp, complete with sentimental piano music and a sexy macho voiceover, and it's at once a revolutionary fist raised on behalf of queers, people of color, the poor, and sex workers. Like Warhol, RuPaul's fascination with fragrance asks the right questions.

Much later, on his wildly successful reality television show *Drag Race,* Ru often offers an advertising challenge. In "Scent of a Drag Queen" during the fifth season, each queen had to create, market and film a commercial for a signature fragrance of their own design. The winner was Alaska's *Red for Filth*, a pun on the need to read the perfume industry for filth, with three deliciously firm critiques functioning as its hilarious tagline: "dangerous, flawless, overpriced."

Marketing says one thing and perfume says another. As far back as 1889, Guerlain created *Jicky* for men but ultimately ended up marketing it for women. Scents have no intrinsic gender or sexuality. There is nothing about this floral note or that amber accord which is inherently discernable as masculine or feminine. Beyond these far from settled debates about who ought to wear what scents, the powdery lemon freshness of *Jicky* was first and foremost artificial. It was the first fragrance to rely on synthetics: the spicy floralcy of linalool, the fresh-cut hay of coumarin, and ethyl vanillin. At the time, this was highly controversial. Synthetics were viewed as self-evidently inferior to naturals, and in some ignorant corners they are still viewed this way. They are different, not inferior, and people who look down their noses at synthetics are missing out on all the flowers that are too delicate to be rendered naturally: freesia, honeysuckle, violet, tulip, gardenia, heliotrope, orchid, lilac and lily of the valley, just for starters. They are missing out on the champagne vibe of *Chanel No. 5* with its overdose of one percent aldehydes, a textural smell of effervescence that has no natural equivalent.

The dumbness of the debate over artificial versus natural ingredients runs parallel to the dumbness of classifying fragrances for men versus for women, and all of this speaks to the dumbness of any authentically gendered selfhood in the first place. Perfume is fluid, flexible, subjective, and possessed of the capacity to perform whatever queer identity one likes. Nowadays, unisex fragrance is all the rage, and hooray for that. It used to be that "to put on airs" literally meant to wear perfume to mask one's own smell until the next available bath time, but eventually the phrase acquired a negative connotation of pretentiousness. Super queer synonyms, slurs and code words for "pretentious" include: arty, affected, conspicuous, extravagant, flamboyant, flashy, flowery, imposing, jazzy, mincing, ornate, vainglorious. And performative.

For me, the briefly transcendent performative capacity of perfume is like Grant Achatz plating my dessert directly on the table at Alinea or Rene Imperato walking the runway with a cane out of time to the soundtrack at the DapperQ fashion show. Oh, one had to be there. These are utterly sublime and perfect creativities in their ugly, fierce, limitless way. One apprehends them immediately and viscerally, as if they are invisibly cloaked—close, very close—by angels all around them. When they draw down the lightning, everything freezes in the scene until it is just them and oneself. One inhales their divine presence, watches them shine with the protection of all those angels, those giants upon whose shoulders we now stand. We walk in fashion. We plate in hospitality. And we in turn become those giants on whose shoulders the future will stand. We, the perfumed, make a home in style. We are surrounded by an ancestral waft of Germaine Cellier, or putting on *Glamazon*, the unisex floral amber powerhouse that RuPaul launched in 2013. One can choose to wear perfume in this manner, as a sign of the magic moment. Or one can wear it as a gender instead.

7 SELF

Nag champa is the most widely known and commonly sold incense in the world. Despite having burned countless sticks of it since my teens and being able to identify it with accuracy in the wild since at least my college years, I have no idea why it took me so many decades to look up what the hell nag champa actually is. It's a mix of sandalwood and a flower, sometimes what Italians call frangipani and Americans call plumeria, but usually Magnolia Champaca. This white flower is generally associated with enlightenment and is often found in Hindu or Buddhist temples. Sometimes this incense also has a dash of some kind of resin or a little spice.

To the untrained nose, nag champa is a basic frame of reference capturing the entire concept of incense without regard for nuances or varietals of any kind. It is a flat smell, perched in the nostril as a single note when a novice is blind to its parts. But even in this elemental apprehension of nag champa, in its straightforward and monotonous glory, its almost message-less arrow can still pierce one's heart. A fragrance can call out to a person, even a person who knows or cares nothing about fragrances. One might grasp a particular scent as having some kinship with the qualities of one's individual identity. We choose a perfume that is an expression of selfhood. Or perhaps we are chosen by it, some part of ourselves called into being by the perfume that speaks to us.

French philosopher Louis Althusser discussed this in the 1970s using a door as a metaphor. When one hears a knock at the door and the voice of a friend, one opens the door. The act of opening the door is simultaneously an act of acknowledging the friendship. One doesn't open the door for a stranger or if there's any sense of danger. The fact that nag champa can compel my unconscious to turn my head means that I am always already a subject who bears some relation to nag champa. I recognize myself in it somehow. The incense hailed me early on and as I grew older, the waft of it from my dorm room door was a signal to other like-minded students that probably we shared a common ideology. And then they'd knock.

A decade or so later, French literary theorist Roland Barthes took Althusser's idea of the hail out of the realm of abstract philosophy and began applying it to the arts. He was interested in the impact a photograph can have on a spectator and classified two kinds of impacts: studium and punctum. Studium is when one's eye sticks to a photo, rather than simply glancing and moving on. It is the objective compositional elements that catch one's attention. For such a long time, I had no notion of how nag champa was made, no ability to understand its composition. And yet it pierced me, punctuated my attention with a pause of genuine attraction. This is the punctum, the sheer feeling that a perfume is a part of one's self, supplanting any cool regard for it as an object of study with the hot immediacy of a wound, of merely loving it.

Because it is an object of study, an engagement with perfume can follow American psychologist Benjamin Bloom's taxonomy of educational objectives. Bloom mapped out a way to approach mastery of any given subject by climbing a kind of pyramid where one achieves more profoundly the higher one moves up this pyramid. He said the basest form of knowledge is simply remembering something, like knowing the ending of a movie. More advanced would be any actual understanding of the movie's ending beyond its basic facts, such as classifying the ending as happy or unhappy. Understanding will breed application, for example relating the movie's ending to the ending of its sequel. Now halfway up the pyramid, the view is beginning to look quite expansive. One begins to analyze how the filmmaker's life story contributed to such an ending for the movie, then perhaps evaluate why this ending makes the movie one's favorite of all time. At the very top of the pyramid lies creativity, where one is inspired to pick up a pen and begin work on a screenplay.

This pyramid model surely applies to the olfactory arts. Our increasing education in life causes our sense of self to evolve, and therefore so too does our scent sensibility. There comes a time in every young perfume lover's life when one finally and quite suddenly begins to smell in three dimensions instead of two. By dimensionality, I mean only a poetic aspect or qualitative feature of fragrance, not the measurable dimensionalities associated with geometry or the molecular level of scent. We've already examined my one-dimensional affiliation with nag champa. This is why so many perfumers scoff at the constant linkage of scent with memory, as if all the magic of what they do is merely a matter of hitting on the right memory. A remembrance is so utterly basic as a form of knowledge.

On Friday, August 12, 1988, my grandmother took me to Six Flags Great America, a theme park filled with roller coasters that I couldn't ride because I was only just about to turn seven. In the gift shop, she bought me a small cedarwood box, thickly shellacked into a beautiful, warm sunset of browns, hinged with brass, a mother and baby unicorn staring at their sparkling reflection in a pool on the box top. I cared nothing for unicorns or boxes, but when I opened the box the smell of that wood and its chemical treatment grabbed me by the throat. I can recall it perfectly. The box cost $7.50 and I still have it, opening it up from time to time to check that my recollection of the scent and the scent itself both remain with relative fidelity. None of the bottles in my perfume collection capture it with precision and if I ever find a bottle that does, no price will be too high.

The scent is not one that I necessarily associate with my grandmother. She introduced me to it by way of the purchase, but I do not love that smell because of her or the box or the special day at the amusement park. I just love it because I loved it. It spoke to me that day without any message and I remember it well. When I was a bit older, my grandmother helped me level up to understand and apply my knowledge. To smell in two dimensions involves synesthesia, the ability to connect what one smells to another perhaps slightly less abstract or at least more familiar thing like a color or an image. When I was maybe ten or twelve, she bought me a Love's scent sampler, their classic drugstore four-pack of plastic splash bottles.

These were the first perfumes that were all mine. The yellow bottle was lemon scented, like kitchen cleaner, easy and literal in the connection between the fruit and its color. The blue one was rain, more on the clean laundry side than the sporty oceanic side; it was easy to associate the cartoonish idea of rain with the color blue. Pink was of course the iconic *Baby Soft* fragrance, a powder fresh scent. I hated pink and seldom used this bottle because its overtly feminizing mission instinctively felt demeaning to the tomboy I knew I was. The fourth bottle was purple. I have no recollection of its fragrance because it had nothing to do with my main purple associations at the time, which would've veered toward either grapes or lilacs. Probably it was a musk, but because the color of it did not correspond to any of my preconceived notions, my brain never quite processed it effectively.

By the time I was in high school, I'd made the move from Love's to Bath & Body Works. B&BW offers a cornucopia of body sprays both classic and seasonal, and at ten bucks for a

236ml plastic bottle it was highly affordable on my limited budget. I quickly landed on a signature scent that I deployed every single day for almost a decade: *Warm Vanilla Sugar*. It's one of their true classics and will likely never be retired, though the formula shifts from time to time. An "oriental" vanilla with some brown sugar on top, a little coconut and rice powder in the middle, generic musks and a gentle sandalwood at the base. Just a basic beige color, unassuming, but an iconic vanilla with tremendous staying power. It was the largest sillage I'd ever encountered.

With *Warm Vanilla Sugar*, I began to apply the ideas of a scent, and to better understand myself as a result. Already by age fifteen, I was a profoundly wounded person—cut down by poverty, by the failures of my parents and the impotence of their rages, by the politics of the Reagan era that I was born into, by a world that wants to take everything from a girl, by social isolation and the consequence-laden oppression of my fledgling queerness. The things that steadied me were banned books, punk rock, the debate team, and the unerring conviction that someday my giftedness would set me free. My open mouth was knife edge, my attitude oozed from every pore. And on top of all this, I swam in a cloud of *Warm Vanilla Sugar*.

This was a scent that was so sweet and so welcoming, so much a snuggly teddy bear, that it stood in clear contrast to all the visual and verbal signaling that I was a big bad monster. It lulled people into a false sense of security before the tongue lashing, advertising my innocence that was then revoked with any word I uttered. It centered, warmed and comforted me so that I could focus on staying alive. It lingered in a room long after I departed, catching everyone in my wide blast radius. It announced me. It was persistent. It left people surprised or confused by the mixed messaging. *Warm Vanilla Sugar* was camouflage and armor at once. It was my business card, my trademark.

But in working like mad to escape my situation in the Nineties, I was in many ways using *Warm Vanilla Sugar* as a cloak of avoidance. Bent on survival, on getting the hell out of Chicago, I made no moves to analyze or evaluate myself too deeply. So badly was I hurting that I felt nothing toward others that I had in turn hurt, ignoring the body count I was racking up so I could save myself. The innate human tenderness inside of me had no room to flourish, and so in retrospect the irony seems clear: during all my youthful perseverance, *Warm Vanilla Sugar* was an anchor to the humanity I could not otherwise afford to show. As I approach middle age, it's impossible to avoid noticing how some of the necessary meanness of my youth has congealed into my character. There are a lot of vanillic gourmand fragrances that appeal to me because I am still chasing my own sweetness.

Since graduate school, since I laid down *Warm Vanilla Sugar*, I have not had a signature scent. As one's fragrance interest levels up on Bloom's pyramid, one hops from bottle to bottle, exercising the faculties of analysis and evaluation with ever quicker and greater precision. When smelling in three dimensions, one truly enters into the scent matrix. One thinks at the level of the molecule but less quantitatively than that, discerning the depth and mood of smells. So Love's *Lemon* is not only yellow, it is solar, cold and upbeat. At some point in the mid to late Nineties, *Warm Vanilla Sugar* once had a seasonal flanker with which I was completely obsessed: *Mulberry Spice*.

At the end of the winter holidays that year, the perfume was discontinued. I didn't have the money or the sense to stockpile it in preparation, and as a result, many times over the past twenty years I have thought about *Mulberry Spice* and how much I loved it. I missed it so much in part because it seemed that I could not ever have it again. It was never my signature scent, but it became one of my grails. A grail fragrance is that bottle for which a scent hound will truly quest. This term refers to the Holy Grail, the Biblical cup that Jesus drank from at his Last Supper and that Joseph then used to collect his blood after the Crucifixion. To hunt for a grail fragrance is to go on a spiritual expedition in search of an enigmatic object of healing that may be more mythic than real.

I'd been checking on eBay for years and finally acquired two bottles of *Mulberry Spice* for about sixty bucks, triple the investment of its sellers, its purple color turned blue with the passage of twenty years. Fortunately, Bath & Body Works uses plentifully cheap synthetic smelly molecules for its body sprays, and so the scent of *Mulberry Spice* is indeed exactly as I remember it. It's not unlike finding a long-forgotten Twinkie in a darkened cabinet corner, where the preservatives have kept it strangely still edible. The difference is that now I have enough perfume education to pick *Mulberry Spice* apart. I didn't realize it was more cinnamon than clove, more musk than vanilla, and my love for it is less now than when I had no ability to analyze it. *Mulberry Spice* was once a revelation to me and having sniffed a hundred cinnamon fragrances since then, I perhaps unfortunately know better now.

Whether the search is for a signature scent or a grail fragrance, it's a lot like playing a game of chess. If one likes Tom Ford's *Oud Wood*, one tries the *Intense* flanker. Then one moves outside the TF collection to other woods, maybe specifically other ouds. One looks at other materials from the same countries of origin or researches the raw materials, moving one square at a time, sometimes forward, other times diagonally. So many strategies in hopes of capturing the queen.

After *Mulberry Spice*, I began nosing around the so-called men's section, then woods. From woods, I realized what I really loved was their smokiness, so I moved to incense fragrances. After learning about incense ingredients, I thought it might not be the smoke itself but the resin. Diving deeply into resins, I spent a long while comparing different kinds and preparations of myrrh. Opopanax proved especially interesting, and from there I sought a cola accord, either a sense of the syrup or the fizz. Many cola accords have a small dose of lemon or orange at the top, so I flung my arms wide around the citrus family from there. Bergamot pointed me toward vetiver and I took an interest in grasses for a while.

One begins to go for samples instead of bottles, paying as much as $7 per milliliter rather than blindly buying the entire $50 or $200 bottle at much lower rates of a dollar or two per milliliter, working quickly through a few or even a few dozen perfumes in a week. One goes around to department stores on weekends to pick up freebies of the latest things, subscribes to newsletters and follows blogs that review the vast numbers of new perfumes launched every season. One keeps lists, takes notes, plans a budget, thinks about storage boxes and filing systems. The urge to compare and contrast these minutiae will rise to the surface for everyone who cultivates a deep interest in perfumery. A person who can smell in three dimensions is ultimately practicing what the Japanese refer to as the Way of Incense. All these tiny leaps, across a single brand or a single note, are a type of *kōdō* game.

Kōdō is the art of appreciating Japanese incense. It is one of the three formal arts of refinement, alongside *kadō* and *chadō*—tea ceremony and flower arrangement. *Kō* is the Chinese word for fragrance. *Dō* is the same as the Chinese *dao* or *tao*, meaning way. The way is both literal because of the incense ceremony and figurative because of the philosophy or ethos of alertness to scent. Agarwood or aloeswood came to Japan from China by way of the Silk Road. Buddhists burned it in the temples, then it became trendy among courtesans and eventually trickled down from elites to regular folks. This all took a thousand years. In the 11th century, in Japan's capital city that we now call Kyoto, Chinese influence was at its peak and noblewoman Murasaki Shikibu wrote one of the first ever novels on planet earth, *The Tale of Genji*.

The story is a very convoluted, aristocratic soap opera, following the life of Genji as he loses and gains political power, has many dangerously secret and scandalously public love affairs, and is generally bored by all facets of living until he dies around age forty. The fifty-four chapters in this psychological novel have gorgeously poetic titles like "The Wind in the Pines" or "The Floating Bridge of Dreams." Fragrance features heavily throughout the tale, and thus the tale itself forms the backbone of formal Japanese incense games. In modern Japan, it is still possible to participate in one of these strict and rule-driven events, but it's mainly considered to be a stuffy pursuit for little old ladies and has faded out of the culture in a way that tea and flowers have not. The complexity of the symbolic novel-based terminology of the games may also be a reason why they never got trendy among the type of Western social groups that are otherwise usually eager to shop a la carte out of the catalog of Eastern rituals.

In the most common *kōdō* game, a master of ceremonies prepares five different packets of incense and marks the back of each packet with the identity of the incense so that she can also participate during the game. She then engages in the ritual ceremony with each packet: lighting a charcoal briquette and burying it halfway down in a pile of white ash inside a hand-held censer, covering it with the ash and then placing a thin mica plate atop the heap, adding a tiny sliver of wood atop the mica plate and tending to it until it heats, then passing the censer around to all participants as the wood becomes fragrant. The participants record what type of incense it is. Sometimes they are trying to match it to an original reference wood chip, other times they are trying to classify it according to the categories of incense first laid out by Sanetaka Sanjonishi, founder of the Oie school of *kōdō*: hot, bitter, sour, salty and sweet. The winner has the highest number of correct identifications.

This game of blind comparing and contrasting is also how apprentice perfumers memorize their materials. Imagine having to correctly identify ten different types of vanilla molecule by fragrance alone—some that are lab-created, others from differing origins across the globe, still others of those same varietals but in differing preparations or strengths of dilution. Imagine having such a specific and detailed scent memory that one can accurately recall each of these ten vanillas without always having to open a sample and sniff. This prerequisite for a career as a perfumer is the fourth and final dimension of any perfume lover's educational progress, wherein one synthesizes concepts to work out a new formula and creates the scent that one has imagined.

Eventually I fixed *Mulberry Spice* by layering it over a more vanillic base fragrance that better brought out the clove note at the top. There was a long detour into the *Ficus* genus that took me from mulberries into figs, until I found myself under the tree where Buddha is said to have obtained enlightenment. Eventually I put a splash of my favorite fragrances into miscellaneous hair and shower products. There was a long detour into the history of laundry detergent manufacturing. Eventually I went back to *Warm Vanilla Sugar*, inserting an opopanax note that I made to give its warmth a less sweet and more sophisticated edge. There was a long detour into safety regulations of the International Fragrance Association to learn about obtaining particular raw materials or perfume extracts that are prohibited for sale in the United States. Eventually I put that same opopanax note underneath Dior's *Poison* in hopes of getting closer to the original formula that went out of production when I was a child, inching toward a scent that I never got to smell in the first place. And then I heard of the Demeter Fragrance Library.

Demeter was founded in 1993 by Christopher Brosius. He was on a mission to bottle everyday scents and launched his first three at New York department stores in 1996: *Grass*, *Dirt* and *Tomato*. As we discussed in the chapter on time, the Demeter scents are single note and their character doesn't shift as the clock ticks. They are meant for layering and wear off very quickly because they contain only as much perfume oil as an aftershave, 2-4% as opposed to an eau de toilette's 5-15% or a perfume extract's 15-40%. Even people who are incompetent with math, people who can't sort out an oil-alcohol-water ratio, can make their own perfumes using the Demeter collection. The first one I made was called *Brunch on the Patio*, and it had yuzu marmalade on top of grass with a mahogany base. I made one called *Astro Church* that was a mix of incense and oud with a spacy metallic note at the opening.

These experiments multiplied like tiny rabbits, covering every sturdy, flat surface of my house with 10ml spray bottles half full of my imagination. The day I ordered five-hundred plastic one-milliliter droppers, I knew for certain I was hooked. I was thinking through new possibilities and variations on previous formulas so quickly that even 10ml creations began to feel too large. After 2ml worth of wear, I could swiftly and easily evaluate, take some notes on how to update the formula and move on to the next. So, to the ongoing delight of all my closest friends and family, I have fallen into a kind of gifting process with these minute experiments once I have had my intellectual fill of whatever is inside each bottle. I'm like Emily Dickinson, circulating her handwritten poems among a small circle of those individuals who took a personal interest in what she was doing, either because what she did was totally killer or because they simply loved her. Some of my creations hit and very many of them do not. I'm a better chef than I am a perfumer, but in time, with practice, who knows.

As my imagination runs the gauntlet of all the fragrance knowledge I have synthesized, squeezing a new formula out into reality every now and then, I've learned to bear in mind the ten virtues of incense associated with this avocation since the 16th century. One, perfume is an aid to meditation. It supports communication with the transcendent.

Two, it purifies the mind and body and three, removes uncleanliness. Four, perfume keeps one alert and five, it provides a companion in the midst of solitude. Six, it brings a moment of peace. It supports tranquility in a busy world. Seven, plentiful use of perfume should never be unpleasant and eight, sparing use of it should still be satisfying. Nine, its aging does not change its potency. And ten, the everyday use of perfume is not harmful.

Searching for a signature scent or hunting down a grail fragrance is an exploration of selfhood. Wearing a perfume is an expression of selfhood. Making a perfume is aspirational. It always points toward a future self where one is a little bit more contemplative, invigorated, attentive and sociable. In Japan, one doesn't "smell" incense. The verb used most regularly to describe apprehending the fragrance is "listening," and in applying this to the matter of asserting one's existence by nose, ye olde tautological Cartesian principle begins to read more like a stunning Buddhist koan: *Olfacio, ergo sum.*

8 OTHER

It is far from easy to gift someone a perfume. There are two main ways of doing it and both of them are terrible. One way is to buy another bottle of whatever that special someone already has at home. Simply refilling a person's stash is insufficient as a gift because the message is only that one has paid attention, not that one has been thoughtful or creative. There's no connectivity implicit in the gesture; it's merely replicative of the special someone's fragrant status quo. One may as well be picking up toilet paper or a dozen eggs. Sometimes it's nice to have an errand taken care of and sure, sometimes a little show of attention is enough of a gift, especially if one has been otherwise negligent recently.

But this kind of gifting also sits on a slippery slope to the clearly awful mode of annual perfume gifting, in which one continually re-ups the scent stash so that the special someone comes to expect and even rely upon the yearly perfume gift. At that point, not only is there no real sense of connection between the two parties, but the special someone is increasingly constrained by the pile up of one bottle so that there's no space remaining for joy or surprise in experimenting or growing. Best to let people stock their own supply of a favorite or a trusted stand-by as needed. The little show of attention implicit in stocking it for them is outweighed by the likelihood of smothering them with it.

The second way is to go to the department store and ask a smiling face with a name tag to please help. More often than not, this is a ten-minute activity. One might state the occasion for the gift, and perhaps the gender expression and occupation of the special someone. Perhaps also name the daily perfume they wear. Then the salesperson will point out two or three bottles and spray them on blotter paper for sniffing. Usually there is light chatter about these bottles being very popular this season, or timeless classics maybe by brands one recognizes. One buys the scent one instinctively likes the most and so leaves the store feeling good about having spent a little time to be thoughtful, to seek some guidance even if it has been quite loose and uninvested.

In this case, we have thoughtfulness divorced from attentiveness. One has given over the process of thinking to the salesperson, who is by nature without any real regard for the special someone. The salesperson is necessary to assist those who are unstudied in how to choose such a gift, but the short duration of the deliberation is a clue to how inadequate this method is. Unless one has stumbled up to the fragrance counter to find a very energetic and deeply caring person behind it, the resultant gift will empty one's pockets without filling up the heart of the special someone. It's the difference between sales and customer service. For two people to find a fragrance that stands a good chance of being suitable to a third, absent person, allot at least an hour.

One should be peppered with questions to give as full a portrait of the special someone as possible because the resultant gift reveals itself through a combination of attention to their unique attributes and thoughtfulness about one's own connection to them. This is regardless of the extent to which the special someone knows anything about perfumery, and is perhaps especially true if they actually don't know anything about it. If the gift is to be a perfume, the giver must be able to explain why this particular bottle was deemed worthy— both suitable and exciting—of the special someone. They can learn about it together and then share in the experience of the special someone wearing it. The feeling of accomplishment in gifting a perfume is located there, in the production of that kind of connectivity, not in the simple satisfaction of plunking the gift box on the table when one gets home.

So let's examine how I arrived at my mother-in-law's seventy-fifth birthday present: a bottle of Etro's *Shantung*. Ellen is known by our family to wear a perfume of roses. If pressed, they might specify that it's a tea rose. This always makes me chuckle—they don't know that most tea rose interpretations smell as much like tea as like flowers, nor that tea roses are some of the least hardy of modern rose types. In the garden, tea roses are known for being high maintenance, so they don't have much in common with my mother-in-law. Ellen is classically sturdy and stoic. She worked as an accounts payable clerk for a Long Island school district while keeping a firm grip on the household management of three kids and a boisterous husband. Photos reveal her keen fashion sense evolving across several decades, remaining both trendy and professional. Even her hairdo always rolled with the times. Yet across the span of fifty years, Ellen is alleged to have worn only this single fragrance of roses.

Naturally, I began digging. The first discovery was that she wears a perfume oil roll-on: *Rose Winds*, by Common Scents. A pun she no doubt still finds cutely witty. It's described as a single note scent that is fresh and romantic, but it's marked neither as pure essential oil nor as a fragrance blend that is exclusive to Common Scents. Thirty milliliters of it costs $35.50, which is a bit much for an oil of such vague provenance, yet extremely thrifty as far how long a single bottle will last. And if she likes the way it smells on her, who cares about all the rest. Still, the randomness of this tiny perfume oil website did not sit right. I couldn't understand how Ellen arrived at it, so I kept digging.

Common Scents is located in Cape Coral, in southwest Florida near Fort Myers, but only since 2006. For the thirty years before that, it was located in Port Jefferson, New York. That's about a fifteen-minute drive from where Ellen raised her three kids. Its owner is Janet DiChiaro, who went to college at St. Joseph's, right near Ellen's house. To the best of Ellen's recollection, they have never met. Yet somewhere in the Seventies, she came to this scent under a highly localized influence, an ad or flyer or a friend of a friend must have launched Ellen in Janet's direction. Then their mutual

Boomerism retired each of them to a corner of Florida where they continue to live out their lives, tied together by this quiet little perfume thread of which neither of them is aware.

Janet's business has one other rose fragrance, *Tea Rose*, which Ellen doesn't use. This led to my second discovery: Ellen uses two rose scents, the other being *Tea Rose*, by Perfumer's Workshop. This is an iconic fragrance, launched in 1977 and sold on the image of Princess Diana, Catherine Deneuve, and Grace Kelly, among other fabulous women. Like my mother-in-law, they are arguably all blondes. *Tea Rose* has sold over fifty million bottles by now and is widely considered a classic. As a trendy thing that became timeless, as an image of womanhood that was strong yet respectable in the eyes of the mainstream, this fragrance very much fits with Ellen.

A 120ml bottle of *Tea Rose* retails for just $19.99. On the top is a quick hit of peony that errs on the side of citrus well enough to be taken for a weak bergamot, and chamomile with a sweetly dry layer that furthers the tea theme while bolstering the fruitishness of the peony through its yellowness. Yet the smell of tea roses themselves is more of a black tea than a yellowish herbal one. A yellow tea is calming, whereas a black one is caffeinating. If the peony is meant to convey a certain energy, then the use of chamomile would seem to have a dampening effect on the refreshing vibe. The smell is clear, but its message is rather mixed.

Perfumer's Workshop says the heart is made up of three roses: tea rose, Damask rose and Bulgarian rose. Tea roses are themselves a hybrid flower and run from white to light pink. The other two roses in this blend are also pink. We're talking about an easy-going, daily kind of femininity here, not the sexy red stuff one breaks out for clinking glasses in the lights of a city at night. Damask roses came from Syria, of course. They no longer grow wild and the cultivated variety is also a hybrid, grown most prolifically in Bulgaria. In the Rose Valley there, the flowers are cut one by one, usually by women. Because each flower possesses very little essence, it takes about 2,000 kilograms of rose petals to produce one kilogram of rose oil.

These flowers are not the shape or color of the rose archetype we envision. These are the roses one grows in the yard, not the ones carefully wrapped in tissue paper to deliver to a special someone. After four centuries and many generations of labor, Bulgarian rose oil is the liquid gold standard because of its healing properties, not its good looks or its great fragrance. It's a working rose, with all kinds of data to support its antibacterial and anti-inflammatory properties, particularly on the skin. I doubt Ellen has been drawn to roses her whole life explicitly because of these medicinal qualities, but she does have vitiligo, a chronic disease where the skin gradually loses blotches of pigment. Perhaps the skin knew what it wanted, and perhaps it told her nose.

In the final dry down, there is the arid filler note of cedarwood, which is also known to be good for skin conditions, though perhaps rosewood would've been equally up to the job of smoothing out the scent while being more to the point of the scent's name. Alongside the evergreen conifer are two kinds of leaves: geranium and violet. These are more green and less peppery, leaving the roses at the heart to provide all the perfume's floralcy, while still carrying the theme of tea by virtue of their dryness and very light spice. The open and close of this perfume are so slight and unassuming that *Tea Rose* really does leave the hodgepodge of roses to do all the work. It's not a scent that most could suss out as Perfumer Workshop's *Tea Rose* in a blind sniff test, but in passing it on the street one could certainly say that there goes a woman who smells like roses.

So it seems that Ellen made up her mind about perfume in the middle of the Seventies, based primarily on those unsought influences that would've landed directly in her lap, filtered through her sensibilities of simplicity and thriftiness. In all, she had a sense of perfume that was not based on any idea of luxuriousness, nor did she pick up an interest in it as a part of the fashion culture to which she was otherwise much attuned. Yet her choices are clearly an expression of self and of a certain notion of serviceability—that if her fragrance isn't broken, there's no need to go out and experiment with fixes to it.

In retirement, Ellen has been somewhat more able to afford luxuries, both financially and psychologically. My in-laws travel, wear some expensive pieces of jewelry, and so on. Even if they've long been able to buy these things, Ellen frequently still falls back on her instinct not to indulge in overspending on flashes in the pan. She is a person who researches things and asks questions, not a person who follows whims or even one who much enjoys experimenting for adventure's sake. She's open to new things, but usually doesn't seek them. For the seeking, she enjoys relying on me and my wife, and we do our level best to deliver. For example, when the four of us make travel plans, it's a running joke that all they need to do is pack a suitcase and wait for me to tell them the itinerary. If we're meeting up in New York City, Ellen will usually suggest a favorite spot from days of yore, but she has never once suggested a place she hasn't previously visited.

The last time we went to Manhattan, we visited one of my wife's favorite stores: the Gucci flagship on Fifth Avenue, where Ellen sniffed their *Alchemist's Garden* line. Of course she was drawn straight to *A Forgotten Rose*, a gorgeously rich oil so haute that the tiny 20ml bottle retails for $420. My guess is that Ellen's total lifetime perfume purchases might barely equal this staggering amount. That's twenty-one bottles of *Tea Rose* or a dozen roll-ons of *Rose Winds*. This fancy Gucci oil, in its resplendent porcelain bottle of sage green with gold accents and a matching box, costs a whopping $21 per milliliter. For the sake of contrast, my personal spending limit in terms of price per milliliter is $7 for an oil and $4 for a spray. The majority of the niche perfumes branch of my collection is 50ml spray bottles in the $200 range.

Unlike the *Rose Winds* roll-on or the generic hunk of rectangular glass housing *Tea Rose*, Gucci's bottle is a work of art in its own right and would be appropriate to display once it is empty. Even their method of display at the store is somewhat unique. They use glass bell jars that sit on scented pads, so one lifts the opening of the jar to the nose. This was most likely the first time Ellen had seen such a posh showcase of fragrance, and in general there is something to be said for the whirlwind of feelings that accompany any trip down Fifth Avenue. One is constantly asked to place oneself on the prestige spectrum through judgments of affordability.

It's a nice unisex scent, deeply redolent of Bulgarian roses with a fresh pop of pink pepper. No origin listed on that pepper, nor on the musk note at the base of the scent. The musk is what carries through the idea of the rose as forgotten, adding a powdery undertow that begins to verge on civet if one dares to hold the bell jar to the nose for too long. In the end, merely ten minutes later, my father-in-law effervesced that he had never smelled anything so wonderful in his life. He declared that Ellen must have this bottle because they could afford such a splurge this once and she would never have considered buying it for herself so that is precisely why she deserved it, and so on. Ellen could see her way to it; she was comfortable with oils and roses already. She would use it for special occasions and make that 20ml last forever.

This bring us to Etro's *Shantung*, which my wife and I gifted Ellen for her seventy-fifth birthday. There is an Etro boutique just a half mile from the Gucci store, and on that same NYC trip, we skipped it to go down to SoHo where the Etro flagship is located. My wife was obviously inspired by her mother's attention to fashion when she was young, and as we all get older, it has been joyful to see how Mindy returns the favor by showing all her favorite shops to her mother. They spent a long time breezing down the racks of neatly hung garments in an array of gorgeously bright colors. I noticed how Ellen kept laying her hands on the sleeves, gauging the softness and weight of each material. She lives in Florida, where silks are only for the sweaty foolish.

And yet, I kept thinking about silk. Because of the slowness and difficulty of asking captive mulberry silkworms to make their cocoons, silk has always been expensive and therefore characterized as regal. But my attention veers toward the workhorse values implicit in the thread—super strong underneath its shine, methodically cultivated over vast spans of time—that make sense as a frame for the infinitude of my mother-in-law's good works. People think of silk as a perfectly smooth surface, but it's not. The most highly prized silks from Italy and China are both slubbed, covered with small visible striations. Slightly more than half of all silk comes from China and there is evidence of its production there as much as 8,500 years ago, long before the trade route running through Silk Road.

Shantung silk is named for Shandong, its province of origin whose name can be literally translated as "east of the mountains." Sericulture, the art of raising silkworms, was long kept a secret there. Confucius, who was also from Shandong, mythologized the invention of silk by telling the story of how the Empress Leizu discovered it when a cocoon suddenly fell into her afternoon tea in the 27th century BC. She fished it out and began to unravel it, soon finding herself standing all the way across the garden from her teacup. So silk was found by a woman's ingenuity, and it was a women's industry for a long time.

This work required focused attentiveness and a good deal of patience. It was not so much a repetitive task as a ritualized one. A ritual is a practice that indeed repeats, but it does so with a careful regard for connectedness. The thread from the cocoon must remain unbroken and then must be woven with many others of similar quality. The process requires not only technical proficiency, but a high degree of respect within its practitioners. As a ritual, silk production is not unlike Ellen's household management to the extent that it relies on a strong foundation of shared values in order to be successful over the long term. As Confucius mused in Book 3 of the *Analects*, the beautiful rituals of painting can only come after plain white silk.

Etro's *Shantung* perfume is meant to capture some sense of this fabric that my mother-in-law actually doesn't wear. And as a gift, determined with deep attentiveness and connectivity, it materially represents a most high respect for Ellen. With its strong roses, the heart note is one she has been loving for many years. With its peony, it carries forward the best part of her old stand-by. With its mandarin momentarily at the top, *Shantung* draws in a reference that connects to my wife, who loves the optimistic energy of orange fragrances above all others. With its black currant note, it gives a nod to the time when Ellen was most cognizant of perfumery and that whiff of cassis was popularized by Guerlain's 1969 perfume *Chamade* as well as Jean-Claude Ellena's 1976 perfume *First* for Van Cleef & Arpels.

But it also pushes at the limit of what Ellen has worn before. The dry down is rounded out by cedar and cashmeran musk. Cedar is my own favorite woody note and one I wear frequently, connecting Ellen's scent to mine. The musk is as warm as that of *Tea Rose*, but it goes in a soft and wooly direction that evokes a sense of physical density more than actual smell. It wraps around the skin's own scent, almost protectively, in a way that is less dry than the base notes of *Tea Rose*. On the whole, it will expand the nuances of her rose vocabulary. Ellen's three other bottles will smell differently in light of whatever she will experience with *Shantung*.

Etro as a brand uses paisley as its trademark. Founded in 1968, the Italian company is not unlike Ellen: priding itself on a kind of new traditionalism, modern and moving with the times, but also highly respectable. *Shantung*'s glass bottle coated in white paisley has more of a heavy hand feel and a substantial presence compared to her two dailies, yet also more whimsy. And as a daily, at $190 for 100ml, it's double the cost per milliliter for *Tea Rose*—a big step up to anytime luxuriousness without tipping over into something so costly she'll be mad at us for overspending. It is a more truly unisex fragrance than Ellen's others and unlike those, it won't be around forever.

Launched in 2016, it's already fading out of Etro's main rotation and bottles of it may soon be somewhat difficult to find. I first sniffed it in 2018, on the trip to New York that Ellen has said stands out in her memory of one of the best trips she's ever taken. My in-laws remember every detail of that long weekend so well and we reminisce about it often. While Mindy and her mother were in Etro's downstairs looking at all the silks, I was upstairs rummaging through the perfume bottles. And I stand by the thought I immediately had upon smelling *Shantung* for the first time: this would suit Ellen very well.

I could not have been more correct. Ellen cherishes this bottle of perfume we got her, perhaps for some of the reasons I've traced here and probably for some more reasons that are entirely her own. One can go on and on in all directions for many pages about any single perfume ever made. For me, there is great pleasure in all the subtleties and minutiae of doing so, noting how perfumes change and how I change with them. To evoke Heraclitus, no one steps into a river of perfume the same way twice. For Ellen, I suspect the truest delight is simply in knowing above all else that I took the time to cogitate so substantively about who she is and what that means to me. This is what we indicate when we say it is the thought that counts. The object is a symbol of the thinking we are doing about each other, connecting us. This particular gift of the sniff transformed us into true friends who deeply understand each other in ways we were previously unable to communicate by any other means. Therein lies the beauty of practicing this more fragrant mode of life: any amount of attention to perfume has the power to change everything.

SELECTED BIBLIOGRAPHY

Achatz and Kokonas, Grant and Nick. *Life, on the Line*. New York: Gotham Books, 2011.

Aftel, Mandy. *Essence and Alchemy: A Natural History of Perfume*. Layton, UT: Gibbs Smith, 2008.

Boisserie, Beatrice, et al. *The Naturals Notebook: Jasmine Sambac*. Paris: NEZ Editions, 2019.

Burr, Chandler. *Emperor of Scent*. New York: Random House, 2004.

Burr, Chandler. *The Perfect Scent*. London: Henry Holt & Co., 2008.

Bushdid, C., et al. "Humans Can Discriminate More than 1 Trillion Olfactory Stimuli." *Science*, 21 Mar 2014: Vol. 343, Issue 6177, pp. 1370–1372.

Chapman, Neil. *Perfume: In Search of Your Signature Scent*. London: Hardie Grant Books, 2019.

Classen, Howes and Synnott; Constance, David and Anthony. *Aroma: The Cultural History of Smell*. London: Routledge, 1994.

Colton, Sara. *Bad Girls Perfume*. Asheville, NC: Water Tower Books, 2016.

Corbin, Alain. *The Foul and the Fragrant: Odor and the French Social Imagination*. Cambridge: Harvard University Press, 1986.

de Bonneval, Eleonore, et al. *The Naturals Notebook: Orange Blossom*. Paris: NEZ Editions, 2020.

de Bonneval, Eleonore, et al. *The Naturals Notebook: Vetiver*. Paris: NEZ Editions, 2020.

de Bonneval, Eleonore, et al. *The Naturals Notebook: Rose*. Paris: NEZ Editions, 2019.

Dore, Jeanne, ed. *The Big Book of Perfume: For an Olfactory Culture*. Paris: NEZ Editions, 2020.

Drobnick, Jim, ed. *The Smell Culture Reader*. London: Routledge, 2006.

Ellena, Jean-Claude. *Perfume: The Alchemy of Scent*. New York: Arcade Publishing, 2011.

Ellena, Jean-Claude. *The Diary of a Nose*. New York: Rizzoli Ex Libris, 2013.

Gilbert, Avery. *What the Nose Knows*. New York: Crown Publishing, 2008.

Groom, Nigel. *The Perfume Handbook*. London: Chapman and Hall, 1992.

Herman, Barbara. *Scent and Subversion: Decoding a Century of Provocative Perfume*. Guilford, CT: Lyons Press, 2013.

Holmes, Bob. *Flavor: The Science of our Most Neglected Sense*. New York: Norton, 2017.

Keller, Helen. "A Neglected Treasure." *The Home Magazine*, June 1934, pp.6. https://www.afb.org/HelenKellerArchive?a=d&d=A-HK02-B225-F02-025.

Keller and Malaspina, Andreas and Dolores. "Hidden consequences of olfactory dysfunction: a patient report." *BMC Ear Nose Throat Disord*, 2013: Vol. 13, Issue 8. https://www.ncbi.nlm.nih.gov/pmc/articles/PMC3733708/.

Laudamiel, Christophe. "Liberte, Egalite, Frangrancite: a fragrance manifesto." 2016. https://american-perfumer.com/blogs/news/liberte-egalite-fragrancite-a-fragrance-manifesto-by-master-perfumer-christophe-laudamiel.

Lauriault, Tracie P. "Scented Cybercartography: Exploring Possibilities." *Cartographica: The International Journal for Geographic Information and Geovisualization*, 2006: Vol. 41, Issue 1, pp.73–91.

McGann, John P. "Poor human olfaction is a 19th-century myth." *Science*, May 2017: Vol. 356, Issue 6338. https://science.sciencemag.org/content/356/6338/eaam7263.

Morita, Kiyoko. *The Book of Incense*. Tokyo: Kodansha International, 1992.

Morris, Matt. "Perfume as Institutional Analysis and Queer Transgression." Presentation to the College Art Association, Los Angeles, 2018. http://www.mattmorrisworks.com/caa-perfume.

Morris, Matt. "Through Smoke and Across Dissent: Power Plays with Perfumery." The Seen: Chicago's *International Journal of Contemporary and Modern Art*, April 2019: Issue 8. https://theseenjournal.org/through-smoke/.

Muchembled, Robert. *Smells: A Cultural History of Odours in Early Modern Times*. New York: Polity Books, 2020.

NEZ: The Olfactory Magazine, Paris: NEZ Editions, volumes 3–10, 2017–2020.

Osman, Ashraf. "Historical Overview of Olfactory Art in the 20th Century." Zurich University of the Arts Seminar Paper, June 2013.

Penny, Louise. *All the Devils are Here*. New York: Minotaur Books, 2020.

Pybus, David. *Kōdō: The Way of Incense*. Boston: Tuttle Publishing, 2001.

Ropion, Dominique. *Aphorisms of a Perfumer*. Paris: NEZ Litterature, 2018.

Roudnitska, Edmond. "Concerning the Circumstances Favorable to the Creation of an Original Perfume." *Perfumer & Flavorist*, April/May 1984: Vol. 9, Issue 2, pp.127–131.

Sanchez and Turin, Tania and Luca. *Perfumes: The A-Z Guide*.
  New York: Viking Penguin, 2008.
Shikibu, Murasaki. *The Tale of Genji*. London: Penguin Classics,
  reprint edition, 2002.
Steingarten, Jeffrey. *The Man Who Ate Everything*. New York:
  Knopf, 1997.
Stoddart, D. Michael. *The Scented Ape*. Cambridge: Cambridge
  University Press, 1990.
Turin, Luca. *The Secret of Scent*. New York: Harper Collins, 2006.
Warren, Cat. *What the Dog Knows*. New York: Simon and Schuster,
  2013.
Westra, Adam. "Nietzsche's Nose." Scents / Sense, Spring 2016,
  pp. 14–17.
Wittig, Monique, translation by Helen Weaver. *The Opopanax*.
  New York: Simon and Schuster, 1966.
Worwood, Valerie Ann. *The Fragrant Mind: Aromatherapy for
  Personality, Mind, Mood, and Emotion*. Novato, CA: New World
  Library, 1996.

INDEX

*Entries stylized as author originally wrote them.

OBJECT LESSONS

Cross them all off your list.

TV

9781501362521

blackface

9781501374012

hyphen

9781501373909

spacecraft

9781501375804

football

9781501367069

perfume

9781501367144

"Perfect for slipping in a pocket and pulling out when life is on hold."
– Toronto Star

9781501358159

9781501358104

9781501348716

9781501353024

9781501344350

9781501361906

Burger by Carol J. Adams

> Based on meticulous, and comprehensive, research, Adams has packed a stunning, gripping expose into these few pages—one that may make you rethink your relationship with this food. Five stars."
>
> *San Francisco Book Review*

> Adams would seem the least likely person to write about hamburgers with her philosophically lurid antipathy to carnivory. But if the point is to deconstruct this iconic all-American meal, then she is the woman for the job."
>
> *Times Higher Education*

> It's tempting to say that *Burger* is a literary meal that fills the reader's need, but that's the essence of Adams' quick, concise, rich exploration of the role this meat (or meatless) patty has played in our lives."
>
> *PopMatters*

High Heel by Summer Brennan

" a kaleidoscopic view of feminine public existence, both wide-ranging and thoughtful."

Jezebel

" Brennan makes the case that high heels are an apt metaphor for the ways in which women have been hobbled in their mobility. She also tackles the relationship between beauty and suffering, highlighting the fraught nature of reclaiming objects defined under patriarchy for feminism."

Paste Magazine

" From Cinderella's glass slippers to Carrie Bradshaw's Manolo Blahniks, Summer Brennan deftly analyzes one of the world's most provocative and sexualized fashion accessories . . . Whether you see high heels as empowering or a submission to patriarchal gender roles (or land somewhere in between), you'll likely never look at a pair the same way again after reading *High Heel*."

Longreads

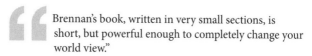

Brennan's book, written in very small sections, is short, but powerful enough to completely change your world view."

Refinery29

In *High Heel*, the wonderful Summer Brennan embraces a slippery, electric conundrum: Does the high heel stand for oppression or power? . . . *High Heel* elevates us, keeps us off balance, and sharpens the point."

The Philadelphia Inquirer

Hood by Alison Kinney

Provocative and highly informative, Alison Kinney's *Hood* considers this seemingly neutral garment accessory and reveals it to be vexed by a long history of violence, from the Grim Reaper to the KKK and beyond—a history we would do well to address, and redress. Readers will never see hoods the same way again."

Sister Helen Prejean, author of
Dead Man Walking

Hood is searing. It describes the historical properties of the hood, but focuses on this object's modern-day connotations. Notably, it dissects the racial fear evoked by young black men in hoodies, as shown by the senseless killings of unarmed black males. It also touches on U.S. service members' use of hoods to mock and torture prisoners at Abu Ghraib. Hoods can represent the (sometimes toxic) power of secret affiliations, from monks to Ku Klux Klan members. And clearly they can also be used by those in power to dehumanize others. In short, *Hood* does an excellent job of unspooling the many faces of hoods."

Book Riot

[*Hood*] is part of a series entitled Object Lessons, which looks at 'the hidden lives of ordinary things' and which are all utterly 'Fridge Brilliant' (defined by TV Tropes as an experience of sudden revelation, like the light coming on when you open a refrigerator door). . . . In many ways *Hood* isn't about hoods at all. It's about what—and who—is under the hood. It's about the hooding, the hooders and the hoodees . . . [and] identity, power and politics. . . . Kinney's book certainly reveals the complex history of the hood in America."

London Review of Books

Personal Stereo by
Rebecca Tuhus-Dubrow

[Rebecca Tuhus-Dubrow's] thoughtfulness imbues this chronicle of a once-modern, now-obsolete device with a mindfulness that isn't often seen in writing about technology."

Pitchfork (named one of *Pitchfork's* favorite books of 2017)

After finishing *Personal Stereo*, I found myself wondering about the secret lives of every object around me, as if each device were whispering, 'Oh, I am much so more than meets the eye' . . . Tuhus-Dubrow is a master researcher and synthesizer. . . . *Personal Stereo* is a joy to read."

Los Angeles Review of Books

Personal Stereo is loving, wise, and exuberant, a moving meditation on nostalgia and obsolescence. Rebecca Tuhus-Dubrow writes as beautifully about Georg Simmel and Allan Bloom as she does about Jane Fonda and Metallica. Now I understand why I still own the taxicab-yellow Walkman my grandmother gave me in 1988."

Nathaniel Rich, author of *Odds Against Tomorrow*

[A] careful, astute study."

The Wire

Souvenir by Rolf Potts

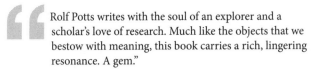

Rolf Potts writes with the soul of an explorer and a
scholar's love of research. Much like the objects that we
bestow with meaning, this book carries a rich, lingering
resonance. A gem."

> Andrew McCarthy, actor, director, and author of
> *The Longest Way Home* (2013)

Souvenir, a sweet new book by Rolf Potts, is a little
gem (easily tucked into a jacket pocket) filled with big
insights . . . *Souvenir* explores our passions for such
possessions and why we are compelled to transport
items from one spot to another."

> *Forbes*

A treasure trove of . . . fascinating deep dives into
the history of travel keepsakes . . . Potts walks us
through the origins of some of the most popular
vacation memorabilia, including postcards and the
still confoundedly ubiquitous souvenir spoons. He
also examines the history of the more somber side
of mementos, those depicting crimes and tragedies.
Overall, the book, as do souvenirs themselves, speaks
to the broader issues of time, memory, adventure, and
nostalgia."

> *The Boston Globe*

Veil by Rafia Zakaria

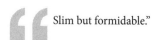

Slim but formidable."

London Review of Books

Rafia Zakaria's *Veil* shifts the balance away from white secular Europe toward the experience of Muslim women, mapping the stereotypical representations of the veil in Western culture and then reflecting, in an intensely personal way, on the many meanings that the veil can have for the people who wear it . . . [*Veil* is] useful and important, providing needed insight and detail to deepen our understanding of how we got here—a necessary step for thinking about whether and how we might be able to move to a better place."

The Nation

An intellectually bracing, beautifully written exploration of an item of clothing all too freighted with meaning."

Molly Crabapple, artist, journalist, and author of *Drawing Blood* (2015)